T0289542

Virtual and Real Labs for Introductory Physics II

Optics, modern physics, and electromagnetism

Virtual and Real Labs for Introductory Physics II

Optics, modern physics, and electromagnetism

Daniel Erenso

Department of Physics and Astronomy, Middle Tennessee State University, Murfreesboro, Tennessee 37132, USA

IOP Publishing, Bristol, UK

ISBN   978-0-7503-3715-1 (ebook)
ISBN   978-0-7503-3713-7 (print)
ISBN   978-0-7503-3716-8 (myPrint)
ISBN   978-0-7503-3714-4 (mobi)

DOI   10.1088/978-0-7503-3715-1

Version: 20211101

IOP ebooks

British Library Cataloguing-in-Publication Data: A catalogue record for this book is available from the British Library.

Published by IOP Publishing, wholly owned by The Institute of Physics, London

IOP Publishing, Temple Circus, Temple Way, Bristol, BS1 6HG, UK

US Office: IOP Publishing, Inc., 190 North Independence Mall West, Suite 601, Philadelphia, PA 19106, USA

I dedicate this book to my father, Bekele Erenso, and my good friend Birhanu Weldemichael. For all academic and non-academic successes, I have achieved in my life, my father, who is no longer alive, takes much of the credit. This book, "Virtual and Real Labs for Introductory Physics II," is built upon my experience in teaching introductory physics courses over the years since my career began in my native country in the early nineties. My good friend Birhanu Weldemichael who is also no longer alive, has contributed to the foundation that helped me land in the USA for my graduate study during the earlier years of my career.

Contents

Preface

Introductory Physics II is the second semester of the algebra or calculus-based sequence of introductory physics courses for science majors. The purpose of this course is to introduce science majors, via virtual and real laboratory activities, to the principles and methods, skills and competencies associated with optics, selected topics in modern physics, and electricity and magnetism. Introductory Physics II is structured to follow Introductory Physics I, which covers selected topics in mechanics, thermodynamics, fluid dynamics, and wave physics. These two courses should provide the student with a broad overview of various introductory physics topics.

The COVID-19 pandemic has created significant challenges for all people in the world. Simultaneously, it has also opened some ample opportunities to overcome these significant challenges of our time. The development of this book was motivated by COVID-19 pandemic challenges. In March 2020, while we were on Spring-break, like all universities in North America and Europe, we moved all ground teaching to remote teaching due to the pandemic. Following this, in my home institution, the university administration had to cancel all summer classes for 2020. To re-open summer classes, the university asked for remote teaching proposals that could get expedite approval. As a professor of physics, I have taught None-calculus-based Introductory Physics II since 2003. In conducting this course, I always revise and update the teaching material by introducing new useful teaching resources that implement technological advances.

The PhET simulation packages developed at the University of Colorado at Boulevard are effective teaching resources developed in recent years. It provides simulation packages that can be applied to create various lab activities that can replicate the real lab in a virtual environment. These PhET simulations packages include Physics, Chemistry, Biology, Mathematics, and Earth Sciences, and it is available in several languages that include Chinese, French, Arabic, Hindi, etc.

This book has nineteen chapters divided into three parts: Optics, Modern physics, and Electromagnetism. Each of the chapters has three sections: a brief introduction to fundamental physical theories, the virtual lab, and the corresponding real lab activities. The virtual lab activities use the PhET simulation packages: Light and Radiation, Quantum Phenomena, Electricity, and Magnets and Circuits. In the following, we list all the simulation packages used for the virtual lab components.

Optics

Optics is the first part of the book. This part of the book covers the reflection and refraction of light, properties of thin lenses, vision correcting lenses, and diffraction of light. Figure 0.1 shows the simulation packages needed for this part of the book.

Modern physics is the second part of this book. In this part, we cover blackbody radiation, the photoelectric effect, introduction to atomic and nuclear physics. The PhET simulation packages shown in figure 0.2 are used for the virtual lab

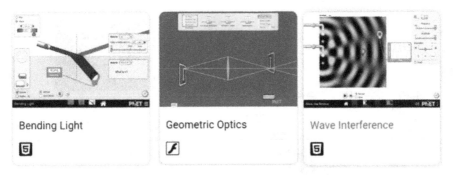

Figure 0.1. Optics PhET simulation packages.

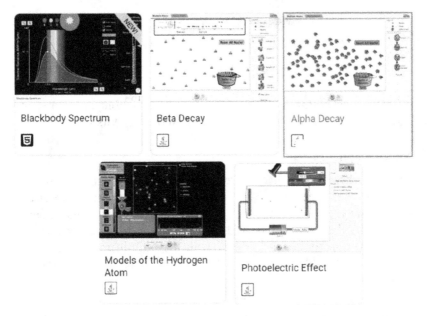

Figure 0.2. Modern physics PhET simulation packages.

activities. Though we have virtual lab components for each topic covered in the modern physics part, the real labs are limited to atomic and nuclear physics sections.

Electricity and magnetism

Electricity and magnetism, which cover both analog and digital electronics, is the last part. For digital electronics, we have only real lab activities. However, for analog electronics, each real lab activity has virtual lab components. Figure 0.3 shows the simulation packages used for virtual labs for this part.

The real labs part uses two PASCO major laboratory kits and a few more pieces of equipment such as a laser pointer, multimeters, ammeters, dc and ac power

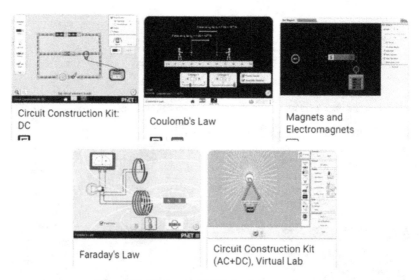

Figure 0.3. Electronics PhET simulation packages.

supplies, function generator, oscilloscope, resistors, capacitors, inductors, and more. We will describe the features of the significant laboratory kits here.

PASCO's basic optics system

This system is easy-to-use, affordable, and ruggedly designed. Large, 50 mm diameter optics components mounted in protective holders that snap directly onto the aluminum track, allowing students to easily adjust components by snapping or sliding them along the track. Image and object distances for both lenses and mirrors can be measured quickly and accurately with the built-in metric tape. The Light Source doubles as a tabletop ray box for studies in reflection, refraction, color addition, and Snell's Law.

Essential physics modular circuits kit

These circuit modules are designed specifically for introductory circuits investigations. For students who have never wired a circuit, this modular system makes it easy for them to see their circuit physically laid-out exactly as it appears in their circuit diagram. Each module connects mechanically to another by sliding the tabs into each other. It works on any tabletop. No special surface is required. To electrically connect two modules, students insert a jumper clip, which emphasizes that an electrical connection has been made. The large size of the modules (8 cm × 8 cm) enables all the students around the table to see and understand the completed circuit.

Finally, this book is designed in a way so instructors can upload each individual virtual or real lab sections as an individual module in their institution platform designed for remote online learning. Furthermore, students can download and write their report in the same pdf file using currently available modern technological devices like iPad and WACOM tablets. Or in cases where such devices are not available, students can simply answer the numerically ordered listed questions by printing on their own paper, scan a PDF file, and uploading to their institution online platforms.

Acknowledgements

Almost all chapters of this book, '*Virtual and Real Labs for Introductory Physics II*', have two major parts. The first is the 'virtual lab', and the second is the 'real lab'. The virtual part is unimaginable without the freely available online simulation packages for physics labs developed by the University of Colorado, at Boulevard, Colorado. The real part is also unthinkable without the equipment I have been using for over a decade, which the Middle Tennessee State University (MTSU) provided. Therefore, I would like to acknowledge the University of Colorado at Boulevard and MTSU.

Author biography

Daniel Erenso

 Dr Daniel Erenso is a professor of physics at Middle Tennessee State University (MTSU), Murfreesboro, Tennessee, USA. He joined MTSU in 2003 after he received his PhD in theoretical physics from the University of Arkansas. Before he came to the USA, Dr Erenso received a BSc (1990) and MSc (1997) in physics from Addis Ababa University in his native country Ethiopia. He also received an Advanced Diploma in Condensed Matter Physics from Abdul Salam International Center for Theoretical Physics (ASICTP), Trieste, Italy, in 1999. For more than two decades, Dr Erenso has served in teaching, research, and mentoring at different universities in his native and adapted counties.

Since he began his service at MTSU, Dr Erenso has taught several introductory and upper-level physics courses. These include algebra-based and calculus-based introductory physics, mathematical methods in theoretical physics, electricity and magnetism, quantum mechanics, and general relativity. For the excellence and dedication that Dr Erenso demonstrated, he has received the MTSU, College of Basic & Applied Sciences Excellence in Teaching award in 2011. More recently, Dr Erenso has demonstrated his dedication and hard work 'by going above and beyond to serve students' during the COVID-19 pandemic.

At MTSU, Dr Erenso has also maintained an active research program with undergraduate students. He has been research advisor for several undergraduate students. His research interest includes theoretical and experimental physics: although by training Dr Erenso is a theoretical physicist in quantum optics since he came to MTSU, he has extended his research to experimental biophysics and quantum optics/quantum information. Dr Erenso has published more than 35, and presented over 60, research works at national and international venues. For his outstanding research accomplishment, Dr Erenso received Sigma Xi the Scientific Research Society Aubrey E Harvey Outstanding Graduate Research Award from the University of Arkansas in 2003, MTSU Foundations Special Project Award in 2005, MTSU, College of Basic & Applied Sciences Distinguished Research Award in 2016 and a nomination for APS Prize for a Faculty Member for Research in an Undergraduate Institution in 2020.

Dr Erenso is a member of several professional societies such as the American Physical Society (APS) and the Optical Society of America (OSA) and serves as an invited reviewer to several international journals. His excellence in teaching and research has also earned him the Fulbright Scholar Award in 2016.

IOP Publishing

Virtual and Real Labs for Introductory Physics II
Optics, modern physics, and electromagnetism
Daniel Erenso

Chapter 1

Reflection and refraction of light

Electromagnetic (EM) waves traveling from one medium to another can be reflected and refracted at the two media's interface. The law of reflection and the law of refraction (Snell's law) describes EM waves' properties at the interface of the two media. This chapter focuses on the verification and application of these laws using virtual and real lab activities. We have also provided a summary of basic theories associated with these fundamental laws. In the virtual lab activity, using the PhTH simulation, we have verified the law of reflection when a laser beam travels from air to glass by measuring the angle of incidence and reflection. Furthermore, by measuring the corresponding angle of refraction (transmission), we have illustrated how one can apply Snell's law to determine the index of refraction for glass experimentally. The virtual lab also demonstrates that when total internal reflection (TIR) can occur, predicting and measuring the critical angle for total internal reflection. These properties are also explored in the real lab activity using a plane mirror and D-shaped glass (cylindrical lens). In both the virtual and the real lab activities, we have used multiple graphical analyses to verify and apply the law of reflection and Snell's law.

1.1 Basic theory

Light
Light is a traveling electromagnetic (EM) wave. It travels with a speed,

$$c = 3 \times 10^8 \text{ m s}^{-1}, \tag{1.1}$$

in vacuum. The speed of EM waves varies from one medium to another. Generally, the speed of a wave v is given by

$$v = \lambda f, \tag{1.2}$$

where f is the frequency and λ is the wavelength. For EM waves, the wavelength, like the speed, varies from one medium to another but not the frequency. The EM wave

doi:10.1088/978-0-7503-3715-1ch1

spectrum consists of a visible and non-visible spectrum grouped by its wavelength. The wavelength range $400 \leqslant \lambda \leqslant 700$ nm is the visible spectrum, commonly referred to as light.

The index of refraction

Suppose an EM wave (light) is traveling with a speed v in a given medium; the refractive index of the medium n is defined by

$$n = \frac{c}{v}, \tag{1.3}$$

where c is the speed of the EM wave in a vacuum. When the EM wave goes from one medium to another, it passes through an interface (or boundary) between the two media. At this boundary, the light could partially be reflected and partially transmitted. There is a change in both the magnitude and direction of the velocity of the transmitted light. This change in the direction of the light as it enters into another medium is called refraction.

The laws of reflection and refraction

Figure 1.1 describes a light across the boundaries of two media with refractive indices n_i and n_t. It shows the line perpendicular to the interface to the two media (the normal), the incident, reflected, and transmitted rays across the boundary. The angle between the normal and the incident ray θ_i is the angle of incidence; the angle between the normal and the reflected ray θ_r is the angle of reflection, and the angle between the normal and the transmitted ray θ_t is the angle of transmission. The law of reflection states

$$\theta_i = \theta_r. \tag{1.4}$$

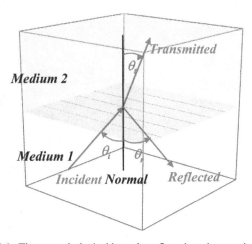

Figure 1.1. The normal, the incident, the reflected, and transmitted rays.

The law of refraction (Snell's law) states

$$n_i \sin(\theta_i) = n_t \sin(\theta_t). \tag{1.5}$$

Total internal reflection and the critical angle (θ_c)

When a light ray crosses the boundary of two media with $n_i > n_t$, the ray bends away from the normal, $\theta_t > \theta_i$. When the angle of incidence increases, the angle of transmission also increases. Total internal reflection (TIR) occurs when the transmission ray is disappearing from the transmitting medium. The angle of incidence beyond which the transmitted light disappears is called the critical angle θ_c. At the critical angle, the transmitted light ray becomes parallel to the interface to the two media (i.e., the transmission angle, $\theta_t = 90°$) and is given by

$$\theta_c = \sin^{-1}\left(\frac{n_t}{n_i}\right), \tag{1.6}$$

which one can derive applying Snell's law.

1.2 Virtual lab: *reflection and refraction*

1.2.1 Introduction

The objectives of this virtual lab are

- To better understand the laws of reflection and refraction of light.
- To experimentally determine the index of refraction of glass using the law of refraction.
- To study what happens to the transmitted light when it goes from less dense to more dense media and vice versa.
- To study total internal reflection and the critical angle.

To this end, go to/click https://phet.colorado.edu/sims/html/bending-light/latest/bending-light_en.html to open the PhET simulation for this virtual lab. You should see a window for the bending light simulation lab shown in figure 1.2. Click the 'More tools' window, and it will direct you to the window shown in figure 1.3. The lower half is the transmitting medium, and it is glass ($n_t = n_g = 1.5$). For the first and second part of this lab, we keep the incident medium air and the transmitting medium glass.

1.2.2 Part I: Law of reflection

In this part of the virtual lab, we are interested in verifying the law of reflection

$$\theta_i = \theta_r. \tag{1.7}$$

We will accomplish this objective by following the procedures listed below.

1. Turn the laser on by clicking the red button. Which direction the transmitted light bends? Away or towards the normal (Note: the vertical dotted line is the normal to the two media interface)? Explain why it bends away or toward the normal.

Figure 1.2. PhET simulation for the laws of reflection and refraction of light.

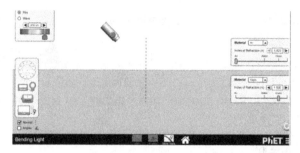

Figure 1.3. The PhET simulation window for the laws of reflection and refraction.

Table 1.1. Table for recording data.

θ_i	θ_r	θ_t	$\sin(\theta_i)$	$\sin(\theta_t)$
0°				
10°				
20°				
30°				
40°				
50°				
60°				
70°				
80°				

2. Drag the protractor from the bottom left corner and set the incidence angle θ_i to zero. What can we say about the angle of reflection θ_r and transmission θ_t when the $\theta_i = 0$? Record the values in the first row at the appropriate columns in table 1.1.
3. Increase the incidence angle by 10°, measure the corresponding angle of reflection and transmission, and record the values at the appropriate corresponding columns in table 1.1. We can change the angle of incidence by dragging the laser in the clockwise or counterclockwise direction.
4. *Law of reflection*: according to the law of reflection, the angle of incidence is equal to the angle of reflection

$$\theta_i = \theta_r. \tag{1.8}$$

5. If we were to plot θ_r vs θ_i, what would you predict for your graph's slope and vertical intercept?

Note that for y vs x graph that displays a linear relationship, the equation is given by

$$y = mx + b, \tag{1.9}$$

where m is the slope and b is the vertical intercept.

6. Using Excel, graph θ_r vs θ_i, and determine the equation for the linear graph. Find the slope and the vertical intercept from the equation found from Excel. Do the values agree with the predicted values in step 5? If not, go back and check the measurements.

On the graph paper provided (figure 1.4), sketch the linear graph you obtained using Excel.

1.2.3 Part II: Snell's law (law of refraction)

1. *Law of refraction*: we now analyze the measurements we took for the incidence and transmission angles. We recall our objective is to determine the refractive index of the glass by applying the law of refraction of light, which states that

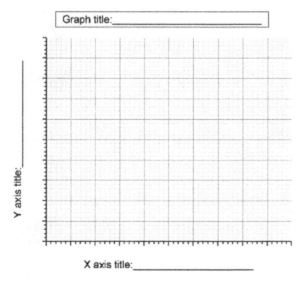

Figure 1.4. Graphing page.

$$n_t \sin(\theta_t) = n_i \sin(\theta_i) \Rightarrow \sin(\theta_t) = \frac{n_i}{n_t} \sin(\theta_i). \tag{1.10}$$

Obviously, from Snell's law, we can see no linear relationship between θ_t and θ_i. In order to get a linear relationship of the form,

$$y = mx + b \tag{1.11}$$

what vs what should we plot? In other words, comparing equations (1.10) and (1.11) what should be y, what should be x, and what should we expect for the slope m and the vertical intercept b?

2. Using Excel make a linear graph and determine the equation for the linear graph.

3. On the graph paper provided (figure 1.5), sketch the linear graph you obtained using Excel.

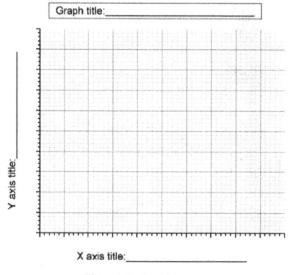

Graph title:_____

Y axis title:____

X axis title:_____

Figure 1.5. Graphing page.

4. Using the linear equation slope find the refraction index for the glass plate n_g.

5. The actual value for the refractive index of glass is $n_g = 1.5$. Calculate the percent difference between the actual value and the experimental value determined in step 4. This percent difference is given by

$$\% \text{ Difference} = \left| \frac{\text{Actual value} - \text{Experimental value}}{\text{Actual value}} \right| \times 100. \qquad (1.12)$$

1.2.4 Part II: Total internal reflection (TIR)

In this part, we are interested in determining the critical angle for a TIR. To this end, we follow the procedure listed below.

1. Set the incidence angle at 45° and change the incident medium to glass and the transmitting medium to water. Which direction does the transmitted light bend? Away or towards the normal?

2. Predict the critical angle θ_c for the glass–water interface.

3. Gradually increase the angle of incidence (θ_i) by rotating the laser in a counterclockwise direction. In the process, we must observe that for $\theta_i > \theta_c$, no light is transmitted into the water, and TIR takes place. What is the measured value for θ_c that you can read from the protractor? Calculate the percent difference between the predicted value and the experimental value for the critical angle.

4. Suppose we change the incident medium to water and the transmitting medium to glass and repeat step 3; do we observe TIR? Explain why?

1.2.5 Result and conclusion

Write a brief overview of what we accomplished and concluded in this activity?

1.3 Real lab: *reflection and refraction*

1.3.1 Objectives

The objectives of this lab are
- To better understand the laws of reflection and refraction of light.
- To experimentally determine the index of refraction of glass using the law of refraction.
- To study what happens to the transmitted light when it goes from less dense to more dense media and vice versa.
- To study total internal reflection and the critical angle.

1.3.2 Supplies

Optics bench, light source, ray-optics table, ray-optics mirror, D-shaped glass (cylindrical lens), slit plate, slit mask, and component mounts (figure 1.6).

1.3.3 Part 1: Reflection

In this part, we shall study the reflection of light from a plane mirror and verify the law of reflection.
1. Affix the ray table, the slit plate, and turn the light-on. The slit plate has seven slits. Align the light passing through the center slit with the 0° mark on the ray table (see figure 1.7).

Figure 1.6. The supplies.

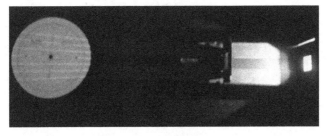

Figure 1.7. The light passing through the seven slits (seven rays).

Figure 1.8. The central ray aligned with 0° mark on the ray table.

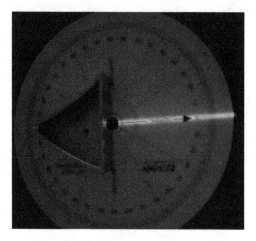

Figure 1.9. The light incident on the plane side of the mirror.

   By moving the ray table closer to and further away from the slits, observe and explain the difference in the rays' properties emerging from the slit plate.

2. Affix the slit mask in front of the slit plate in order to block all the light rays except the central ray (see figure 1.8).

   (Note: Make sure the ray is further away from the source as much as possible and the central ray is lined-up with the 0° mark on the ray table.)

3. Place the ray-optics mirror at the center on the ray table with its plane reflecting surfacing facing the light (see figure 1.9).

4. Rotate the ray table with an increment of 10° up to 80° (see, for example, figure 1.10 for $\theta_i = 10^0$) and record the angle of incidence θ_i and angle of reflection θ_r at the appropriate columns in table 1.2.

5. Make a ray diagram that shows the mirror reflecting surface, the normal, the incident, the reflected rays, and the corresponding angles.

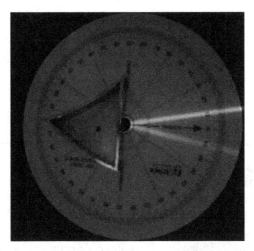

Figure 1.10. The incident and reflected ray for angle of incidence θ_i.

Table 1.2. Data table.

θ_i	θ_r	θ_t	$\sin(\theta_i)$	$\sin(\theta_t)$
0°				

6. *Law of reflection*: according to the law of reflection the angle of incidence is equal to the angle of reflection,

$$\theta_i = \theta_r.$$

If we were to plot θ_r vs θ_i, what would you predict for your graph's slope and vertical intercept?

7. Using Excel, graph θ_r vs θ_i, and determine the linear graph equation. Do the slope and the equation's intercept determined from Excel agree with the predicted slopes and intercepts in step 6? If not, go back and check the measured value.

8. On the graph paper provided (figure 1.11), sketch the linear graph you obtained using Excel.

9. Next, we shall collect the data that we need to study refraction law in the next part. To this end, we replace the ray-optics mirror with the D-shaped glass (the cylindrical lens), but we keep everything else the same. In aligning the cylindrical lens, follow the steps listed below.

    (a) Align the central ray with the 0° mark on the ray table.
    (b) Align the plane side of the cylindrical lens with the ray table's diameter (the line passing through the 90° mark). This side of the cylindrical lens must also face towards the light source such that the central ray passes through the center of the cylinder (see figure 1.12).

10. Increase the angle of incidence θ_i by 10° up to 80° (see for example figure 1.13 for $\theta_i = 10^0$) and record the angle of transmission θ_t at the appropriate columns in table 1.2.

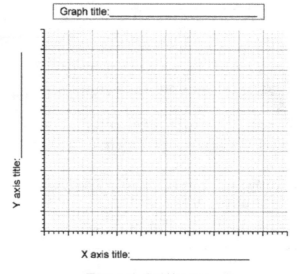

Figure 1.11. Graphing page.

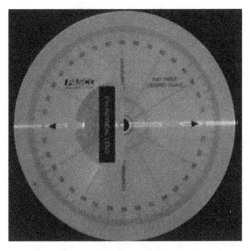

Figure 1.12. The incident (in air) ray and the transmitted (into glass) ray, for an angle of incidence $\theta_i = 0°$.

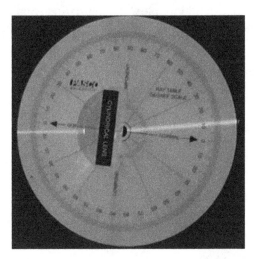

Figure 1.13. The incident (in air) ray and the transmitted (into glass) ray, for an angle of incidence $\theta_i = 10°$.

11. Make a ray diagram that shows the cylindrical lens, the normal, the incident and transmitted rays, and the corresponding angles at the lens's two surfaces.

1.3.4 Part II: Law of refraction and TIR

1. Using Excel, graph θ_t vs θ_i for your data in table 1.2.
2. Sketch your graph for θ_t vs θ_i in figure 1.14. What is the difference between the θ_r vs θ_i and θ_t vs θ_i graphs?
3. *Law of refraction*: we now proceed to analyze the measurements we took for the angle of incidence and transmission angle. We recall our objective is to determine the refractive index of the glass by applying the law of refraction of light, which states that

$$n_t \sin(\theta_t) = n_i \sin(\theta_i) \Rightarrow \sin(\theta_t) = \frac{n_i}{n_t} \sin(\theta_i). \tag{1.13}$$

Obviously, from Snell's law, we can see no linear relationship between θ_t and θ_i. In order to get a linear relationship of the form

$$y = mx + b, \tag{1.14}$$

what verses what should we plot? By comparing equations (1.13) and (1.14), determine the relationship between the slope m and the refractive indices?

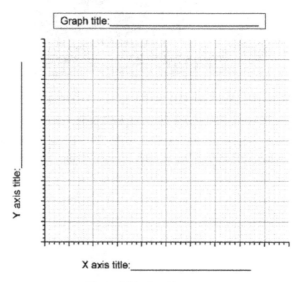

Figure 1.14. Graphing page.

4. Using Excel, make a linear graph, and determine the equation for the linear graph.

5. Using the relationship determined in step 3 and the result in step 4, calculate the index of refraction for the glass n_g.

6. The actual value for the refractive index of glass is, $n_g = 1.5$. Calculate the percent difference between the actual and the experimental value determined in step 5,

$$\text{Difference} = \left| \frac{\text{Actual value} - \text{Experimental value}}{\text{Actual value}} \right| \times 100. \qquad (1.15)$$

7. We now collect the data that we need to study total internal reflection (TIR). We use the same cylindrical lens but a slightly different procedure. We keep the central ray lined-up with the 0° mark on the ray table and the cylindrical lens's plane side lined-up with the ray table's vertical diameter (the line passing through the 90 ° mark). But this time, the curved side of the lens must face towards the light source such that the central ray passes through the center of the cylinder (see figure 1.15).

8. Our goal here is to study TIR. We already know that TIR takes place when the angle of incidence is greater than the critical angle θ_C which is the angle of incidence where the corresponding angle of transmission $\theta_t = 90°$. It is given by

$$\theta_c = \sin^{-1}\left(\frac{n_t}{n_i}\right).$$

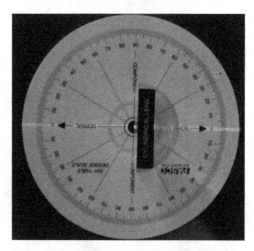

Figure 1.15. The curved surface of the lens is facing the light source. The incident ray (in the glass) and the transmitted ray (in the air), for angle of incidence $\theta_i = 0°$.

Using the experimentally determined value for the refractive index of the cylindrical lens (glass) and the refractive index for air, find θ_c.

9. Rotate the ray table with an increment of 5° or less (when necessary) until you reach the critical angle θ_c. Record the angle of incidence θ_i and the corresponding angle of transmission θ_t in table 1.3 (see for example figure 1.16 for $\theta_i = 10°$).

10. What is the critical angle observed in step 9? Find the percent difference between this value and the predicted value in step 8? Does the percent difference show the experimental and predicted values agree within the limit of experimental uncertainties? What would happen when the angle of incidence exceeds the critical angle?

11. Make a ray diagram that shows the two surfaces of the cylindrical lens, the normal, the incident and transmitted rays, and the angles of incidence and transmission at the two surfaces.

Table 1.3. Data table.

θ_i	0°										
θ_t											

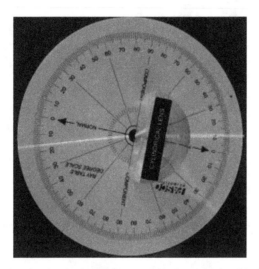

Figure 1.16. The incident ray (in the glass) and the transmitted ray (in the air), for an angle of incidence $\theta_i = 10°$.

12. Use Excel to graph θ_t verses θ_i for your data in table 1.3. Sketch this graph and the two graphs you found for θ_r vs θ_i and θ_t vs θ_i for the data in table 1.1 (figure 1.14) in figure 1.17. Describe the similarities and differences between these three graphs.

13. By facing the two sides (plane and curved) of the cylindrical lens, we have studied the refraction of light from air into the glass and from the glass into air media. In the sketch shown in figure 1.17, you must have shown one linear and two non-linear graphs. Label each of the non-linear graphs as air to glass or glass to air and explain how we knew that.

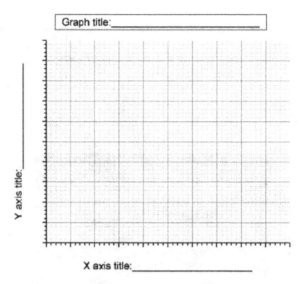

Figure 1.17. Graphing page.

1.3.5 Result and conclusion

Write a brief overview of what we have accomplished and concluded in this activity.

IOP Publishing

Virtual and Real Labs for Introductory Physics II

Optics, modern physics, and electromagnetism
Daniel Erenso

Chapter 2

Thin lenses

In chapter 1, we have studied the refraction of light as it travels from one medium to another. In this chapter, we study how we locate and describe images formed due to the refraction of light passing through thin lenses. The thin lens equation and a ray diagram are the two methods that generally are used to locate and describe images formed by thin lenses. This chapter provides a summary of the fundamental theories, simulation labs that use the PhTH and The Physics Classroom simulation packages, and a real lab that uses a PASCO optics kit. In both the simulation and the real labs, a set of instructions for the measurement of the object and the corresponding image distances are given. Students are also instructed to analyze the measured data graphically and answer conceptual questions associated with locating and describing images formed by thin lenses, in particular a converging lens.

2.1 Basic theory

Focal point
A focal point is where a parallel beam of light rays converges to, or appears diverging from, after it undergoes refraction from an object. Objects that are capable of causing such kind of properties are called lenses (for example, see figure 2.1).

Focal distance (F)
Focal distance is the distance from the focal point to the center of the lens.

Thin lenses
A lens is a thin lens when the distance from one side of the lens to the other is much smaller than the focal distance. We will consider two types of lenses: converging (convex) and diverging (concave) lenses (see figure 2.2). The focal length f for a converging lens is equal to the lens's focal distance (i.e., $f = F$) and a diverging lens $f = -F$.

doi:10.1088/978-0-7503-3715-1ch2

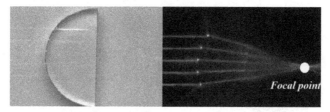

Figure 2.1. The focal point for a cylindrical lens.

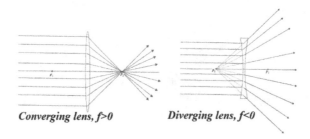

Converging lens, f>0   *Diverging lens, f<0*

Figure 2.2. Converging and diverging lenses.

Lens image description

There are two ways to locate and describe an object's image in a thin lens. The first method is the analytical method, which uses the thin lens equation:

$$\frac{1}{f} = \frac{1}{d_o} + \frac{1}{d_i},\tag{2.1}$$

where d_o is object position and d_i is image position. This equation relates the focal length to the object and image positions. If we know any two of these quantities, we can determine the third using this equation. While object and image distances are always positive, object and image positions can be positive or negative.

$$d_0 = \begin{cases} \text{positive} & \text{(Object is real)} \\ \text{negative} & \text{(Object is virtual)} \end{cases}\tag{2.2}$$

$$d_i = \begin{cases} \text{positive} & \text{(Image is real)} \\ \text{negative} & \text{(Image is virtual)} \end{cases}\tag{2.3}$$

The magnification of an image m determined by

$$m = -\frac{d_i}{d_0} = \frac{h_i}{h_o}\tag{2.4}$$

is used to describe the orientation and the size of an image relative to the object. Note that h_i is the height of the image and h_o is the height of the object.

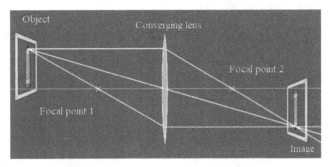

Figure 2.3. A ray diagram for an object placed at $d_o > f$.

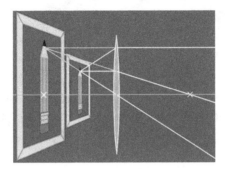

Figure 2.4. A ray diagram for an object placed at $d_o < f$.

$$m = \begin{cases} \text{positive,} & \text{image is upright} \\ \text{negative,} & \text{image is inverted} \end{cases} \tag{2.5}$$

The $|m|$ describes how bigger or smaller the image is relative to the object.

$$\begin{cases} |m| > 1, & \text{image is magnified} \\ |m| < 1, & \text{image is minified} \\ |m| = 1, & \text{image is same as the object} \end{cases} \tag{2.6}$$

The second method for locating and describing images by thin lenses is by making a ray diagram. In a ray diagram, we use three principal rays drawn to scale from the object to locate and describe the image accurately.

    (a) The ray leaving the object's tip and traveling parallel to the optical axis will pass through the focal point F_2 after passing through the lens.

    (b) The ray leaving the object's tip and passing through the focal point F_1 will emerge from the lens, traveling parallel to the optical axis.

    (c) The ray leaving the object's tip and passing through the center of the lens will emerge from the lens undeflected.

Figure 2.3 shows the three principal rays drawn to scale. It shows the image is real and inverted when the object distance is greater than the focal distance. On the other hand, when the object distance is less than the focal distance, the ray diagram in figure 2.4

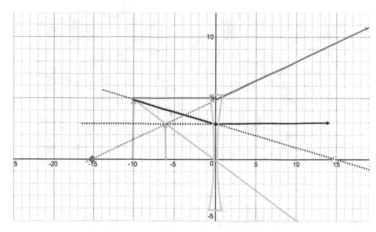

Figure 2.5. The three principal rays for a diverging lens used to locate the image.

shows that image is virtual, upright, and magnified. Figure 2.5 shows the ray diagram for an object sitting in front of a diverging lens. We used the three principal rays to locate and describe the image formed. Independent of the object's position, the image formed by a diverging lens is always virtual, upright, and minified.

2.2 Virtual lab A: converging *lens*

2.2.1 Introduction

The objectives of this virtual lab are
- To better understand thin lenses and the properties of these lenses associated with object and image distances.
- To locate and describe images formed by thin lenses using the thin lens equation and a ray diagram.
- To determine the focal length of thin converging lens, applying the thin lens equation.

To this end, we open the PhET simulation for this virtual lab by going to/clicking https://phet.colorado.edu/en/simulation/legacy/geometric-optics. You should see a window for the geometrical optics simulation lab shown in figure 2.6.

Note: Do not click on any of the boxes or icons unless you are told to do so! If you do, please reopen the simulation.

2.2.2 Part I: an object and a thin lens

1. Is the lens a diverging or converging lens?

2. Click on the 'Show help' box. How many focal points does the lens have?

3. Is the focal length of this lens positive or negative?

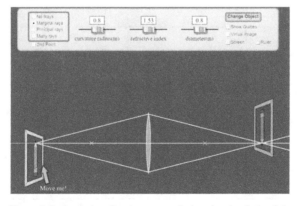

Figure 2.6. A window for the geometrical optics simulation lab.

4. Is this image real or virtual?

5. Is this image upright or inverted?

6. Does the image look magnified, minified, or about the same size as the object?

7. Click on the 'Ruler' box, move the ruler close to the lens, and measure the focal distance. What is the focal length f of the lens?

8. Set the object (the pencil) position (d_0) at 120 cm and measure the corresponding image position (d_i). Record the values at the appropriate row and column in table 2.1. Change the pencil's position by dragging it close to or away from the lens.

   Note: Make sure that the lens's curvature radius, refractive index, and diameter are kept at the default values shown in figure 2.7

Table 2.1. Table for recording data.

d_o (cm)	d_i (cm)	$\frac{1}{d_O}\left(\frac{1}{cm}\right)$	$\frac{1}{d_i}\left(\frac{1}{cm}\right)$
120			
130			
140			
150			
160			
170			
180			

Figure 2.7. Default setting for the curvature of the lens, the refractive index of the lens, and diameter of the lens.

9. Increase the object position by 10 cm and measure the corresponding image position until it reaches 180 cm. Record the values at the appropriate rows and columns in table 2.1.

10. In the thin lens equation,

$$\frac{1}{f} = \frac{1}{d_o} + \frac{1}{d_i} \Rightarrow \frac{1}{d_i} = \frac{1}{d_o} - \frac{1}{f}, \tag{2.7}$$

d_i and d_o do not show a linear relationship. To get a linear relationship of the form

$$y = mx + b, \tag{2.8}$$

what vs what should we plot? In other words, comparing equations (2.8) and (2.7) what should be y, what should be x, what should be the value of the slope m? What is the relationship between the focal length f and the vertical intercept b?

11. Using Excel, make a linear graph and determine the equation. To make the linear graph, calculate $1/d_o$ and $1/d_i$ and record the values in table 2.1.

12. Applying the relationship between the intercept and the focal length determined in step 10, find the focal length of the lens, f.

13. On the graph paper provided (figure 2.8), sketch the linear graph you obtained using Excel in step 11.

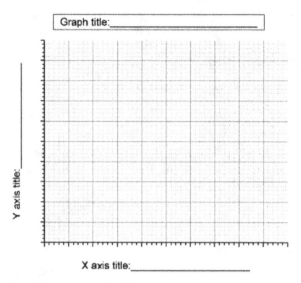

Figure 2.8. Graphing page.

14. The actual value for the focal length of the lens is what you measured in step 7 and the measured value is the focal length you determined applying the thin lens equation in step 12. Calculate the percent difference between the actual and the measured values,

$$\% \text{ Difference} = \left| \frac{\text{Actual value} - \text{Measured value}}{\text{Actual value}} \right| \times 100. \qquad (2.9)$$

2.2.3 Part II: questions

1. In our description of thin lenses, we have seen how to locate and describe an image formed by a thin lens by tracing three principal rays. Re-position the object at 180 cm away from the center of the lens and click the icon 'Principal rays'. (a) Describe the properties of the three principal rays that you see. (b) Based on what you see from the ray diagram, describe the image formed by this lens.

2. Keep the default value for the lens's curvature radius, refractive index, and diameter (see figure 2.7). Position the object at a distance less than the focal length of the lens. What happens to the principal rays traced from the pencil tip after it crosses the lens?

3. Click on the box 'virtual image', move the object back and forth between the lens and its focal point, and describe the image formed by a converging lens in such a particular case.

4. Re-position the object at 120 cm and keep the lens's curvature radius, refractive index, and diameter the same at the default value shown in figure 2.7. Discuss what happened to the lens or the image when (a) the lens's curvature radius, (b) the lens's refractive index, and (c) the lens's diameter changes.

2.2.4 Result and conclusion

Write a brief overview of what accomplished and concluded in this activity?

2.3 Virtual lab B: *converging*

2.3.1 Introduction

The objectives of this virtual lab are

- To better understand thin lenses and the properties of these lenses to object and image distances.
- To locate and describe images formed by thin lenses using the thin lens equation and a ray diagram.
- To determine the focal length of thin converging lens, applying the thin lens equation.

To this end, we open the Physics Classroom simulation for this virtual lab by going to/clicking on https://www.physicsclassroom.com/Physics-Interactives/Refraction-and-Lenses/Optics-Bench/Optics-Bench-Refraction-Interactive. Scroll down to find the Optic bench interactive simulation window. There is a small hot spot in the top-left corner. Click on the hot spot to open the interactive in full-screen mode.

2.3.2 Part I: an object and a converging lens

1. The default setting for the Optic bench interactive simulation lab is the converging lens. Ensure that the window shows a ray diagram for an object (a candle) sitting in front of a converging lens.

2. How many focal points does the lens have?

3. Is the focal length of this lens positive or negative?

4. Is this image real or virtual?

5. Is this image upright or inverted?

6. Does the image look magnified, minified, or about the same size as the object?

Table 2.2. Data read from the simulation window.

f (cm)	d_o (cm)	d_i (cm)	h_o (cm)	h_i(cm)

7. Read the focal length f, the object distance d_o, image distance d_i, object height h_o, and image height h_i and record the values in table 2.2.

8. Determine the focal length f from the thin lens equation

$$\frac{1}{f} = \frac{1}{d_o} + \frac{1}{d_i}.$$ (2.10)

Does the result agree with the value for f recorded in table 2.2?

9. Calculate the magnification using the values recorded in table 2.2

$$m = -\frac{d_i}{d_0},$$ (2.11)

and

$$m = \frac{h_i}{h_o}.$$ (2.12)

Does the result confirm the answer given in steps 4–6?

10. Set the object (the candle) position (d_0) at 30 cm and read the corresponding image distance (d_i). Record the values at the appropriate row and column in table 2.3. To change the object's distance, drag the object close to or away from the lens.

Table 2.3. Data read and calculated.

d_o (cm)	d_i (cm)	$\frac{1}{d_o}\left(\frac{1}{\text{cm}}\right)$	$\frac{1}{d_i}\left(\frac{1}{\text{cm}}\right)$
30			
40			
50			
60			
70			
80			

11. Increase the object distance by 10 cm up to 80 cm and read the corresponding image distance. Record the values at the appropriate rows and columns
in table 2.3.

12. Obviously, in the thin lens equation,

$$\frac{1}{f} = \frac{1}{d_o} + \frac{1}{d_i} \Rightarrow \frac{1}{d_i} = \frac{1}{d_o} - \frac{1}{f},$$ (2.13)

d_i and d_o do not show a linear relationship. To get a linear relationship of
the form

$$y = mx + b,$$ (2.14)

what vs what should we plot? In other words, comparing equations (2.14)
and (2.13) what should be y, what should be x, what should be the value of
the slope m? What is the relationship between the focal length f and the
vertical intercept b?

13. Using Excel, make a linear graph and determine the equation. To make the
linear graph, calculate $1/d_o$ and $1/d_i$ and record the values in table 2.2.

14. Applying the relationship between the intercept and the focal length, find the lens's focal length f.

15. On the graph paper provided (figure 2.9) sketch the linear graph, you obtained using Excel in step 13.

16. The actual value for the lens's focal length is what you read in step 7, and the experimental value is the focal length you determined in step 14. Calculate the percent difference between the two.

$$\% \text{ Difference} = \left| \frac{\text{Actual value} - \text{Measured value}}{\text{Actual value}} \right| \times 100 \qquad (2.15)$$

17. Re-position the object at 80 cm away from the center of the lens. (a) Describe the properties of the three principal rays that you see. (b) Based on what you see from the ray diagram, describe the image formed by the lens.

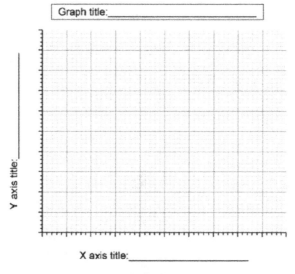

Figure 2.9. Graphing page.

18. Position the object at a distance less than the focal length of the lens. What happens to the rays traced from the candle tip after it crosses the lens?

19. Position the object at a distance half the focal length from the lens and describe the image formed by a converging lens in such a particular case.

2.3.3 Result and conclusion

Write a brief overview of what you accomplished and concluded in this activity.

2.4 Real lab: *thin lens equation*

2.4.1 Objectives

The objectives of this lab are
- To study the properties of converging and diverging lenses and these lenses' behavior associated with object and image distances.
- To determine the focal length of a thin converging lens, applying the thin lens equation.

2.4.2 Supplies

Optics bench, light source, three lenses, crossed arrow target, viewing screen, component mounts (figure 2.10).

2.4.3 Procedure

Part I: estimating the focal length of a thin lens
There are three lenses for this lab. Two of the lenses are converging lens, and the third is a diverging lens. This activity aims to verify the focal lengths of these lenses using a simple procedure that we can derive from the thin lens equation. If we put the object far away, like at infinity ($d_o \simeq \infty$), from a lens with a focal length f, where would be the image position? Well, from the thin lens equation,

$$\frac{1}{f} = \frac{1}{d_i} + \frac{1}{d_0} = \frac{1}{d_i} + \frac{1}{\infty} = \frac{1}{d_i} \Rightarrow \frac{1}{f} = \frac{1}{d_i} \Rightarrow f = d_i, \qquad (2.16)$$

where we used $1/\infty = 0$. For a distant object, the image position is at the focal point of the lens.

1. Using the lenses provided, try to project a clear and sharp image of any distant object inside or outside the room on a white screen or a white paper by moving the lens or the screen closer or apart.

2. Measure the distance between the lens and the screen where the object's clear and sharp image has been formed.

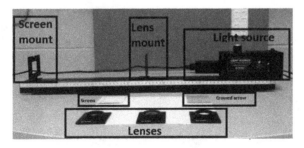

Figure 2.10. The supplies.

Table 2.4. Measured and advertised focal lengths and experimental the uncertainties.

Lens type	Measured focal length (cm)	Uncertainties (cm)	Advertised focal length (cm)

3. Record the measured values with the corresponding uncertainties and the advertised values for the focal lengths in table 2.4.

4. Among the three lenses, there is one where there is no image projected on the screen, and therefore, it is impossible to measure the focal length using such a procedure. Explain why.

Part II: measuring the focal length of thin converging lens
In the virtual lab activity, by measuring the object's and the corresponding image's positions and conducting a graphical analysis, we have determined the focal length for a converging lens applying the thin lens equation

$$\frac{1}{f} = \frac{1}{d_i} + \frac{1}{d_0}. \tag{2.17}$$

We will do a similar analysis. To this end, we follow the following procedure:
1. Affix the Crossed Arrow Target directly to the light source and the lens with focal length 7.5 cm (750 mm) to the lens mount. Position the crossed arrow target at the 10 cm mark and the lens at the 35 cm mark on the optical bench.

2. Affix the Viewing Screen onto the screen mount and position it on the opposite side of the lens to the light source. Adjust the position until a clear and sharp image of the object (the crossed arrow) is projected on the screen.

3. Find the object distance (d_o) and image distance (d_i) from the optical bench and record the values in table 2.5.

4. Move the object closer to the lens by 2 cm, and the image gets blurry. Adjust the position of the screen to get a sharp image and repeat step 3.

Table 2.5. Measured object and image distances.

d_o(cm)	d_i (cm)	$\frac{1}{d_o}\left(\frac{1}{cm}\right)$	$\frac{1}{d_i}\left(\frac{1}{cm}\right)$

5. Repeat step 4 five times.

6. Calculate $\frac{1}{d_o}$ and $\frac{1}{d_i}$ and record the values at the appropriate columns in table 2.5.

7. Estimate the uncertainties for the measurements: Δd_o and Δd_i.

8. Calculate the average values for object distance $(\bar{d}_o)$ and image distance $(\bar{d}_i)$.

9. Calculate the fractional uncertainties for object distance (FU$_o$ = $\frac{\Delta d_o}{\bar{d}_o}$) and image distance (FU$_i$ = $\frac{\Delta d_i}{\bar{d}_i}$).

2.4.4 Data analyses

We now proceed to analyze the image and object positions. We recall our objective is to determine the focal length of the converging lens using the thin lens equation

$$\frac{1}{f} = \frac{1}{d_i} + \frac{1}{d_0} \Rightarrow \frac{1}{d_i} = -\frac{1}{d_0} + \frac{1}{f}. \tag{2.18}$$

1. From this equation we can see that there is no linear relationship between d_i and d_o. In order to get a linear graph of the form

$$y = mx + b, \tag{2.19}$$

what vs what should we plot? That means what should be y, what should be x, what should be the intercept b, and what should we expect for the value of the slope m comparing equation (2.19) to the thin lens equation in equation (2.18)?

2. Using Excel, make a linear graph and determine the equation for the linear graph. Sketch the linear graph obtained on the graph paper provided in figure 2.11.

2.4.5 Result and conclusion

1. Write the equation of the best-fit straight line for the linear plot in step 2. Use the appropriate symbols for the x and y axes titles in figure 2.11.

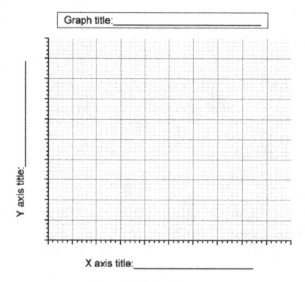

Figure 2.11. Graphing page.

2. What is the value of the lens's focal length f obtained from the linear equation?

3. Determine the fractional uncertainty for the focal length. Use the highest fractional uncertainty in the measured values.

4. Express the experimental value for the focal length with the uncertainties. Does the experimental value agree with the advertised value within the limit of uncertainties? If not, explain why.

5. Write a brief overview of what was accomplished and concluded in this activity.

IOP Publishing

Virtual and Real Labs for Introductory Physics II
Optics, modern physics, and electromagnetism
Daniel Erenso

Chapter 3

Human eye and corrective lenses

Chapter 2 studied the thin lens equation to locate and describe images formed by converging and diverging lenses. Optics of the human eye is an exciting topic in physics as it illustrates the thin lens equation's application as light travels through the eye. A person's eye with normal vision can focus as close as 25 cm (near point) and as far as infinity (far point). When the near point is greater than 25 cm, the person is said to have a vision defect known as hyperopia (farsightedness). When the near point is less than infinity, the person is said to have a vision defect known as myopia (nearsightedness). Lenses can correct such vision defects, and this chapter discusses applying the thin lens equation to determine the power of such corrective lenses for hyperopia and myopia. It begins with a basic description of the human eye's optics and these two associated vision defects. It then provides a set of procedures in a real lab environment to determine corrective lenses' power, commonly prescribed in diopters.

3.1 Basic theory

Normal eye
The cornea, the lens (a converging lens), the ciliary body, the pupil, the retina, the vitreous humor, and the optic nerve all play a major role in human vision. When a normal human eye focuses on an object, a light diverging from the object is incident on the eye and undergoes refraction through the cornea, the lens, the vitreous humor resulting in the formation of a sharp image on the retina. The optic nerves detect this image and send a signal to the brain, and we would be able to see the object clearly. A normal eye can see as far as infinity and as close as 25 cm. The farthest distance we can see is known as the far point (D_{max}), and the closest distance we can see is known as the near point (D_{min}). For a normal eye,

$$D_{max} = \infty, \ D_{min} = 25 \text{ cm} \tag{3.1}$$

Figure 3.1 models when a normal eye (lens 1) focuses on an object, a clear and sharp image is formed on the retina. The eye creates the image of the object we are looking at on the retina by accommodating the lens with the help of the ciliary muscles. It means the ciliary muscles on the two opposite edges of the lens increase the focal point by pulling the lens outward and making it thinner (less curved) or decrease the focal length by pushing the lens inward and making it thicker (more curved). When the ciliary muscle fails to do this task, we get a vision problem. We will focus on two common vision problems in the human eye: farsightedness (hyperopia) and nearsightedness (myopia).

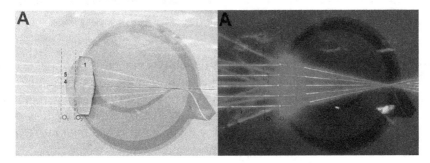

Figure 3.1. A normal eye model lens (lens 1). It focuses light on the retina.

Hyperopia (farsightedness)

Hyperopia is a vision defect in an eye with difficulties to see objects at the normal near point. When the object is placed at the normal near point, a hyperopic model eye lens (lens 3), shown in figure 3.2, focuses the light coming from the object beyond the retina, and the image becomes blurry. For a hyperopic eye, the near point is greater than the near point for a normal eye,

$$\text{Hyperopic eye, } D_{\min} > 25 \text{ cm.} \tag{3.2}$$

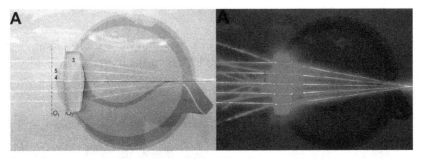

Figure 3.2. A hyperopic eye. The image of an object placed at the normal near point is formed beyond the retina.

Hyperopia can be corrected by a converging lens with the appropriate focal length. Figure 3.3 shows when the right corrective converging lens (lens 4) is placed in front of a hyperopic model eye lens (lens 3), the light coming from the object is focused on the retina.

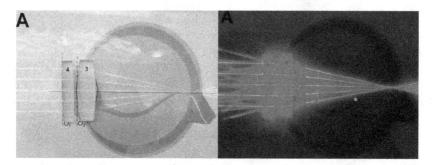

Figure 3.3. A model for a hyperopic eye corrected by a converging lens.

Myopia (nearsightedness)

Myopia is a vision defect when an eye has difficulty seeing at the normal far point. For a myopic eye, light coming from an object placed at the normal far point, which is at infinity, is focused by the model eye lens (lens 2) before the retina, as shown in figure 3.4. Such a vision defect can be corrected by wearing a diverging lens with the appropriate focal length. The far point for a myopic eye is less than infinity,

$$\text{Myopic eye,} \quad D_{\max} < \infty. \tag{3.3}$$

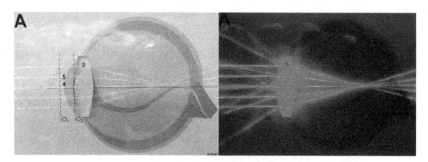

Figure 3.4. A model for a myopic eye corrected by a converging lens.

Figure 3.5 shows a model corrective diverging lens with the right focal length (lens 5) in front of a model myopic eye lens (lens 2) focussing the light from an object placed at a normal far point.

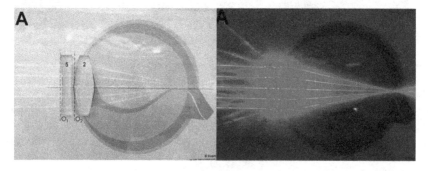

Figure 3.5. A myopic model eye (lens 2) corrected by a diverging lens with the right focal length (lens 5).

Lens maker equation
The focal length f of a lens is determined by the lens maker equation,

$$\frac{1}{f} = \left(\frac{n_l}{n_m} - 1\right)\left(\frac{1}{r_1} - \frac{1}{r_2}\right), \tag{3.4}$$

where r_1 and r_2 are the radii of curvature of the front and back surfaces of the lens, respectively, n_l is the refractive index of the glass that the lens is made of, and n_m the refractive index of the medium that the lens is in. After crossing the lens's surface, if the light passes through the region where the center of curvature of that surface is located, the radius of curvature is positive. Otherwise, the radius of curvature is negative.

Dioptic power
The power P of corrective lenses prescribed for vision defects is measured by the focal length of the lens

$$P = \frac{1}{f}, \tag{3.5}$$

where the focal length f is measured in meters. The unit is diopter (dp), $1 dp = 1/m$.

3.2 Real lab: corrective *lenses*

3.2.1 Objectives

The objectives of this lab are
- To better visualize the two common vision defects (hyperopia and myopia) in a lab setting.
- To determine the dioptic power of corrective lenses for hyperopia and myopia, applying the thin lens equation.

3.2.2 Supplies

The supplies for this activity, for the most part, are the same as the supplies used for the real lab activity in chapter 2. These include: optics bench, light source, three lenses, crossed arrow target, viewing screen, component mounts except for additional second lens mount that we use to mount the corrective lens (see figure 3.6).

3.2.3 Procedure

This activity aims to learn how to determine the dioptic power of corrective lenses for hyperopia and myopia by applying the thin lens equation. The two lenses ($f = 150$ mm and $f = -150$ mm) are the corrective lenses that we are interested in determining the dioptic power experimentally. To this end, we first identify for what vision defect these two lenses can be prescribed for, calculate the dioptic power,

$$P = \frac{1}{f},$$
(3.6)

and record the values in table 3.1.

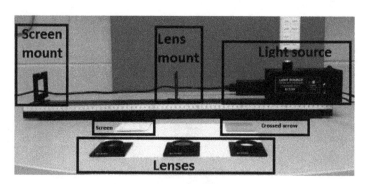

Figure 3.6. The supplies.

Table 3.1. Calculated dioptic power for the corrective lenses.

Lens type	Prescribed for	Advertised f (m)	Calculated DP (dp)

Part I: hyperopia

For a normal eye, the near point is 25 cm from the eye. For a hyperopic eye, the near point is always greater than 25 cm. Let us consider a person with hyperopia, whose near point is 100 cm. We are interested in correcting this vision defect using a corrective lens. Using the right corrective lens, we want this person to see objects (like a person with normal vision) placed at 25 cm away from the person's eye. The corrective lens with the right dioptic power allows the person to see by creating a virtual image at the hyperopic person near point (100 cm) for the real object placed 25 cm away from the eye. We have made our prediction that can correct this vision defect for this individual in table 3.1. In order to verify our prediction, we will proceed as follows. We first model the retina by the viewing screen, the eye's lens by the converging lens with $f = 7.5$ cm (75 mm), and what the person is trying to see by the crossed arrow target. The 25 cm near point translates to the 40 cm mark on the optical bench, and the 100 cm near point translates to the 60 cm mark.

1. Affix the crossed arrow target directly to the light source and place it at the 60 cm mark on the optical bench.

2. Affix the viewing screen onto the screen mount and position the screen at the 0 cm mark on the optical bench.

3. Affix the eye model lens with focal length 7.5 cm to the lens mount and adjust the position until a crossed arrow target's sharp image is visible on the screen.

4. Move the object to the 40 cm mark on the optical bench. What happens to the image on the screen (on the retina)? At this position, the object's sharp and clear image form before or after the viewing screen (retina)? (Remember, we are dealing with hyperopic eye here.)

5. Affix the right corrective lens to the second lens mount. Place this lens on the optical bench between the 7.5 cm focal length lens (the eye model lens) and the object (crossed arrow target) find the best possible position that would allow you to see a sharp image on the screen (retina).

6. Determine the object distance (d_0), the virtual image distance (d_i) that the hyperopic eye sees (i.e., the near point for the hyperopic eye on the optical bench). We measure these distances from the corrective lens position on the optical bench. Record the values for d_0 and d_i, the corresponding

uncertainties, Δd_0 and Δd_i, the calculated fractional uncertainties, FU_{d_0} and FU_{d_i} on table 3.2. Should the image distance d_i be negative or positive? Do not forget that the image created by the corrective lens is virtual. People wearing corrective glasses never see the actual object; instead, they see the object's virtual image created by the corrective lens at their near point.

Table 3.2. Object and image distances with the corresponding uncertainties.

d_0 (cm)	d_i (cm)	Δd_0 (cm)	Δd_i (cm)	$FU_{d_0} = \frac{\Delta d_0}{d_0}$	$FU_{d_i} = \frac{\Delta d_i}{d_i}$

Part II: myopia

For a normal eye, the far point is at infinity ($=\infty$) from the eye. For a myopic eye, the far point is always less than infinity. Let us consider a person with myopic eye that cannot see beyond 300 cm from his eye. We are interested in correcting this vision defect using a corrective lens with the right dioptic power for the person's eye to see objects placed at infinity, as a person with normal vision. The right corrective lens takes an object placed at infinity and creates a virtual image at the myopic person's far point (= 300 cm). In table 3.1, we have calculated the dioptic power of this lens. In order to verify this result, we will follow a similar procedure we followed in Part I. We model the retina by the viewing screen, the eye lens by the converging lens with $f = 7.5$ cm, and the object that the person tries to see by the crossed arrow target. Unlike Part I, however, on the optical bench, the far point for a normal eye ($=\infty$) translates to the farthest distance, the 60 cm mark, and the far point for the myopic person (300 cm) translates to the 40 cm mark.

1. Remove the first corrective lens from the second lens mount but keep the screen's (0 cm mark) and the crossed arrow target (40 cm mark) at the same positions. Adjust the eye model lens (i.e., $f = 7.5$ cm) until you see a sharp image on the screen (on the retina).

2. Move the object back to the 60 cm mark on the optical bench. What happens to the image on the screen (retina)? At this position, the clear and sharp image for the object forms before or after the screen (retina)? (Remember, we are dealing with myopic eye here.)

3. Affix the right corrective lens to the second lens mount and adjust its position till a clear and sharp image is visible on the screen (retina).

4. Determine the object distance (d_0), the virtual image distance (d_i) that the Myopic eye sees (i.e., the far point for the myopic eye on the optical bench). We measure these distances from the corrective lens position on the optical bench. Record the values for d_0 and d_i, the corresponding uncertainties, Δd_0 and Δd_i, the calculated fractional uncertainties, FU_{d_0} and FU_{d_i} on table 3.3.

Should the image distance d_i be negative or positive? Do not forget that the image created by the corrective lens is virtual. Again, people wearing corrective glasses never see the actual object; instead, they see the object's virtual image created by the corrective lens at their far point, in this case.

Table 3.3. Object and image distances with the corresponding uncertainties.

d_0 (cm)	d_i (cm)	Δd_0 (cm)	Δd_i (cm)	$FU_{d_0} = \frac{\Delta d_0}{d_0}$	$FU_{d_i} = \frac{\Delta d_i}{d_i}$

Table 3.4. Experimental result for the two corrective lenses.

Vision defect	f (m)	DP (dp)

3.2.4 Data analyses

1. Using the measured values for the image and object position (tables 3.2 and 3.3) and the thin lens equation

$$\frac{1}{f} = \frac{1}{d_i} + \frac{1}{d_0},$$
(3.7)

   calculate the focal length f and then the dioptic power ($DP = 1/f$, where f is in meters). Also, use the highest fractional uncertainties to estimate the focal length and dioptic power uncertainties. Record the results in table 3.4.

   Do the experimentally calculated results in table 3.4 agree with the values recorded in table 3.1, within the limit of uncertainties?

2. Suppose the lenses used in this lab ($f_1 = 7.5$ cm, $f_2 = 15$ cm, and $f_3 = -15$ cm) were made from glass with refractive index $n_l = 2.5$. The focal lengths' values determined are in the air with refractive index $n_m = 1.0$. Find the radius of curvature for each lens. Both surfaces of the lenses have the same radius of curvature. Make schematics showing the two surfaces and the corresponding center of curvature for each lens.

3.2.5 Result and conclusion

Write a brief overview of what has been accomplished and concluded in this lab activity.

IOP Publishing

Virtual and Real Labs for Introductory Physics II
Optics, modern physics, and electromagnetism
Daniel Erenso

Chapter 4

Diffraction and gratings

This chapter introduces you to the properties of light associated to its wave nature when it passes through a narrow aperture or slits. This property of light is called diffraction, which is a result of interference observed when the diameter of the aperture or the distance between the slits d is comparable to the wavelength λ (i.e. $d \sim \lambda$). We will consider a rectangular plate of glass that has a number of slits per unit length known as a *diffraction grating* and a *circular aperture*.

4.1 Basic theory

Two slits interference
Consider two slits separated by a distance d. The slits are at a distance L from a screen, as shown in figure 4.1. When a plane monochromatic light of wavelength λ shines on these slits it undergoes an interference resulting in alternating bright and dark fringes when the slit width and slits separation is comparable to the wavelength of light. The interference fringe spacing Δy projected on the screen in figure 4.1 is given by

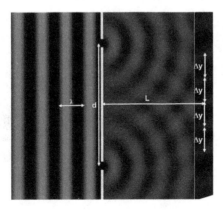

Figure 4.1. Two slits interference.

doi:10.1088/978-0-7503-3715-1ch4

Figure 4.2. Diffraction grating.

$$\Delta y = \frac{\lambda L}{d}. \tag{4.1}$$

Diffraction grating

A grating is usually made by etching many equally spaced slits or 'lines' on a glass sheet (see figure 4.2).

If the number of lines per unit length is n, the distance between adjacent slits (see figure 4.1) is given by

$$d = \frac{1}{n}. \tag{4.2}$$

If this distance is comparable to the wavelength of a monochromatic light λ, the light incident on the grating when viewed on a distant screen, an alternating pattern of bright and dark spots is observed as a result of interference. The bright region is due to the constructive and dark region is due to the destructive interference of the light incident on the grating. Figure 4.3 illustrates the diffraction of monochromatic red light of wavelength, $\lambda = 632\,\text{nm}$, from a laser pointer. The diffraction equation describes the diffraction pattern:

$$d \sin(\theta) = m\lambda, \tag{4.3}$$

Figure 4.3. Diffraction of a monochromatic red light.

where d is the distance between the adjacent lines in the grating, $m = 0, 1, 2, ...$ is the order of diffraction, and θ is the angle between the mth order and the zero-order of diffraction, and λ is the wavelength. In spectroscopy, a diffraction grating is useful for separating the colors in an incident white light (see figure 4.4).

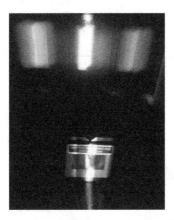

Figure 4.4. Diffraction of a white light by a grating.

Diffraction by circular aperture
When a monochromatic light of wavelength λ, shines through a tiny circular aperture of diameter, d comparable to λ it results in different interference fringes shown in figure 4.5. These interference fringes consist of a bright spot at the center surrounded by alternating dark and bright rings. We are interested in the first and the second dark rings for the diffraction pattern of monochromatic light by a circular aperture. The radius of the first dark ring r_1 defines what is known as the Airy disk, which is given by

$$r_1 = \frac{1.22\lambda L}{dn}, \qquad (4.4)$$

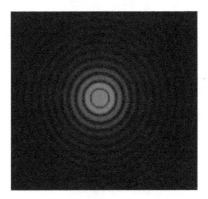

Figure 4.5. Diffraction of a red light ($\lambda = 632$ nm) from a laser pointer by a circular aperture with diameter d.

and the radius of the second dark ring r_2

$$r_2 = \frac{2.23\lambda L}{dn},$$

(4.5)

where d is the circular aperture diameter, L is the screen's distance from the circular aperture, and n is the medium's refractive index between the screen and the circular aperture (see figure 4.6).

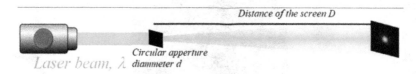

Figure 4.6. A schematic for diffraction of monochromatic light by a circular aperture.

4.2 Virtual lab I: *two slits*

4.2.1 Introduction

The objectives of this virtual lab are

- To study the interference pattern of a monochromatic light wave in double slits.
- To determine the wavelength for a monochromatic light by analyzing this light's interference pattern in double slits.
- To study the parameters that determine the separation distance between the interference fringes in double slits.

To open the simulation for this lab go to/click https://phet.colorado.edu/en/simulation/wave-interference. You will see the simulation window shown in figure 4.7. Click play, and it opens another window shown in figure 4.8. In this activity, as our goal is to study interference in a double slit, click on 'Slits,' and it directs you to the simulation lab window shown in figure 4.9.

Figure 4.7. Wave interference PhTH simulation lab.

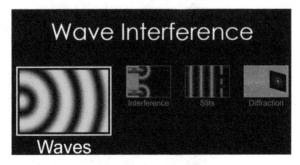

Figure 4.8. Wave interference simulation labs.

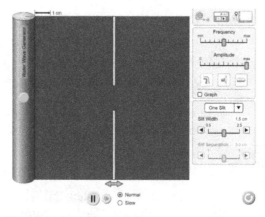

Figure 4.9. Simulation window for slits wave interference.

4.2.2 Part I: fringe vs slits separation

In this part of the lab, we want to determine a green laser's wavelength by measuring adjacent fringes' distance vs the distance between the two slits. To this end, we follow the procedure listed below.

1. Click on the symbol for a laser pointer (the third symbol on the fourth row on the right-hand side).

2. The default setting is for slit number (one slit), slits width (500 nm). Change the slit number from one to two. The default slits separation distance is 1500 nm. Keep the slit width the same (500 nm) but change the slit's separation distance to 1000 nm.

3. Check the 'screen' and 'intensity' boxes and turn on the laser. The default wavelength (frequency) setting for the laser pointer is green. Keep this wavelength throughout this part of the activity.

4. Using the tape measure (located on the first row on the right-hand side), measure the screen's distance from the slits plate L and record the value.
   For additional help, if it is necessary, see figure 4.10.

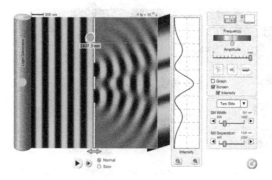

Figure 4.10. Measuring the distance of the screen from the slits plate.

5. By keeping the screen distance the same and setting the slits separation distance, d, to the values listed in table 4.1, measure the corresponding fringe separation distance Δy. Wait a half or one minute after changing the silt separation distance to see the interference pattern on the screen.

6. The interference fringe spacing Δy in double slits is related to the slits separation d by the equation,

$$\Delta y = \frac{\lambda L}{d},\qquad(4.6)$$

where L is the distance of the screen from the slits plate and λ is the wavelength of the incident light. There is no linear relationship between Δy and d. Compare equation (4.6) with the generic linear equation,

$$y = mx + b,\qquad(4.7)$$

and identify the x and y variables, the slope m, and the vertical intercept b that gives a linear graph.

Table 4.1. Data for Δy vs d.

Slits separation d (nm)	$1/d$ (nm)	Fringe separation Δy (nm)
1000		
1200		
1400		
1600		
1800		
2000		
2200		
2400		
2600		
2800		

7. In step 6, we should identify $1/d$ and Δy as x and y variables, respectively. Calculate $1/d$ and record the values in table 4.1. Create a plot in Excel for Δy vs $1/d$ for the data in table 4.1, and have Excel draw a best-fit straight line for the data and display the equation of the line. As always, watch the symbols and units in the equation for the best-fit straight line. Then in the space provided in figure 4.11 sketch your plot.

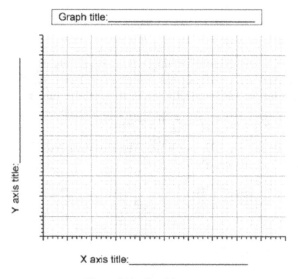

Graph title:_____

Y axis title:_____

X axis title:_____

Figure 4.11. Graphing page.

8. Write down the equation for the best-fit straight line.

9. Use the slope's value from the equation in step 8 and the slope's expression in step 6 and determine the incident light's wavelength.

10. Move the slits plate as far as possible away from the laser source (see figure 4.12).

11. Measure the distance from the first to the last bright (green) stripes. Count the number of bright stripes and determine the wavelength of the light

Figure 4.12. The wavefront for the light incident on the slit plate.

incident on the slits plate. Be careful about how this distance, the number of bright stripes, and the wavelength are related.

12. What is the percent difference in the results for the wavelength determined in steps 9 and 11? What could be the possible source of uncertainties that contributed to the differences in the two results?

4.2.3 Part II: fringe spacing vs screen distance

In this part of the activity, we want to study the interference fringes spacing as we change the distance of the screen from the slits plate. This time we use a red laser instead of a green laser.

1. Change the laser from green to red. Keep everything else the same.

2. Position the slits plate about 1000 nm from the laser source. It may be a little bit hard to set the distance to 1000 nm precisely. Any distance that is close to this value is good enough to proceed to the next step.

3. Measure the distance from the first to the last bright (red) stripes that you see between the laser source and the slits plate. Count the number of bright stripes and determine the wavelength of the light incident on the slits plate.

4. By increasing the distance of the slits plate from the screen up to 3000 nm by an amount convenient, record the measured values for L and Δy in table 4.2 (at least get 5-10 pairs of data).

Table 4.2. Data for Δy vs L.

Screen distance L (nm)	Fringe separation Δy (nm)

5. Create a plot in Excel of Δy vs L for the data in table 4.2, and have Excel draw a best-fit straight line and display the equation. Write down the equation and sketch the graph in the space provided in figure 4.13.

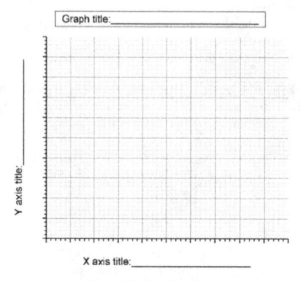

Figure 4.13. Graphing page.

6. Use the value of the slope in the previous question to determine the wavelength of the laser.

7. Find the percent difference between the experimental values in steps 3 and 6. Does the percent difference indicate the two results agree within the experimental uncertainties?

4.2.4 Result and conclusion

1. The three parameters that determine the interference fringes spacing Δy are the spacing between the slits d, the screen's distance from the slits plate, and the wavelength. Keep any of the two parameters and change the third and observe the interference pattern on the screen. Discuss the observation for each case.

2. Write a brief overview of what has been accomplished and concluded in this activity.

4.3 Virtual lab II: *circular aperture*

4.3.1 Introduction

The objectives of this virtual lab are
- To study the diffraction pattern resulting from the interference of mono-chromatic light waves passing through a circular aperture.
- To better understand the Airy disk and its size changes as the diameter of the circular aperture changes.
- To carry out linear graphical analyses from data we generate by measuring the diameters of the aperture and the corresponding radii of the Airy disk.

We recall that when a monochromatic light with wavelength λ passes through a circular aperture with diameter d, we see an interference pattern shown in figure 4.6 on a screen placed at a distance L behind the circular aperture. A closer look at the circular aperture and the interference pattern on the screen is shown in figure 4.14(a) and (b). The interference pattern shows a green bright spot followed by alternating dark and bright green rings. The central bright spot surrounded by the first dark ring is the Airy disk. The Airy disk has a size determined by the radius of the first dark ring surrounding the central bright spot, which is shown by the white circle in figure 4.14 (c). This radius of the Airy disk, denoted by r_1 is given by

$$r_1 = \frac{1.22\lambda L}{dn},\qquad(4.8)$$

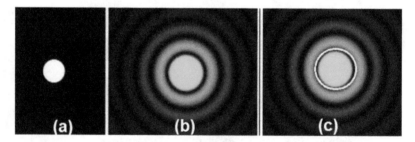

Figure 4.14. (a) The circular aperture, (b) the interference pattern, and (c) the size of the Airy disk (the white circle).

where n is the refractive index of the medium between the aperture and the screen.

In this virtual lab, we will have a better understanding of the diffraction of light by a circular aperture by examining the relationship between the size of the Airy disk r_1, the wavelength, λ, and the diameter of the aperture d. To this end, go to/click http://phet.colorado.edu/sims/html/wave-interference/latest/wave-interference_en.html to open the PhET simulation lab. A window (see figure 4.7) for the wave interference lab we used for two slits will open up. Click play, and another window we used in the virtual lab for two slits will also open up (see figure 4.8). We are interested in the interference of light as it passes through a circular aperture. Select the Diffraction, and it will open the window shown in figure 4.15.

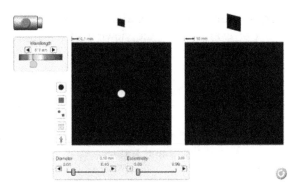

Figure 4.15. Diffraction of light virtual lab for different shape of apertures.

4.3.2 Finding the distance of the screen, L

1. This experiment's default setting is $\lambda = 511$ nm (green laser), a circular aperture with diameter $d = 0.1$ mm zero eccentricity. Do not change the eccentricity of the aperture throughout the experiment.

2. Turn the laser on by clicking the red button on the top-left corner to see the circular aperture's interference pattern discussed earlier. There is a length scale on the top sides of the magnified view for the aperture and the screen. Record the length scale for the screen, which is 10 mm. Measure the corresponding 'scaled length' l_{scaled} using a ruler from your viewing device. Calculate the scaling factor SF

$$SF = \frac{l_{actual}}{l_{scaled}} = \frac{10 \text{ mm}}{l_{scaled}}. \tag{4.9}$$

For example, for my viewing screen,

$$l_{scaled} = 280 \text{ mm} \tag{4.10}$$

and the scaling factor becomes

$$SF = \frac{10 \text{ mm}}{l_{scaled}} = 0.036. \tag{4.11}$$

3. Do not zoom in or zoom out the viewing screen once the scaling factor sets to the value calculated in step 2. Such action could skew the next measurements as this scaling factor is what we will be using through this virtual lab.

4. Using a ruler, measure the scaled diameter for the Airy disk and calculate the radius r_1. For example, for the viewing screen I am using, the scaled diameter of the Airy disk $d_{A,\,scaled}$ is

$$d_{A,\,scaled} = 390 \text{ mm} \Rightarrow r_{1,\,scaled} = \frac{d_{A,\,scaled}}{2} = 195 \text{ mm}. \tag{4.12}$$

5. Using the scaling factor determined in step 2 and the Airy disk's scaled radius from step 3, find the actual radius $r_{1,\,actual}$.

For example, using the scaling factor and the scaled radius calculated above, one finds the actual radius for the Airy disk to be

$$SF = \frac{r_{1,\,actual}}{r_{1,\,scaled}} \Rightarrow r_{1,\,actual} = SF \times r_{1,\,scaled} = 7.02 \text{ mm}. \tag{4.13}$$

6. Record the values for the diameter of the circular aperture d, scaled radius $r_{1,\,scaled}$ and the actual radius $r_{1,\,actual}$ for the Airy disk on table 4.3.

Table 4.3. Measured and calculated data for a diffraction from a circular aperture.

$d(mm)$	d_i (mm)	$r_{1,\,scaled}$ (mm)	$r_{1,\,actual}$ (mm)	$\frac{1}{d}\left(\frac{1}{mm}\right)$
0.06				
0.10				
0.14				
0.18				
0.22				
0.26				
0.30				

7. Repeat steps 4 to 6 for the circular aperture diameter listed in table 4.3. Click the right or left arrows near the bottom of the simulation window to set the circular aperture diameter to these values.

8. We recall the equation for the radius of the Airy disk,

$$r_1 = \frac{1.22\lambda L}{dn}. \tag{4.14}$$

Since this virtual lab simulates an experiment carried in a laboratory room, the medium is air, and we set $n = 1$ so that

$$r_1 = (1.22\lambda L)\frac{1}{d}. \tag{4.15}$$

9. We can see no linear relationship between r_1 and d. In order to get a linear relationship of the form

$$y = mx + b, \tag{4.16}$$

what vs what should we plot? In other words, comparing equations (4.15) and (4.15) what should be y, what should be x, what should be the value of the slope m? Express the distance of the screen from the aperture L in terms of the slope m?

10. Calculate $1/d$ and record the values in the appropriate column in table 4.3.

11. Using Excel, make a linear graph and determine the equation for the best-fitting straight line.

12. On the graph paper provided (figure 4.16), sketch the linear graph you obtained using Excel.

13. Using the equation for L determined in step 9, the laser's wavelength in the experiment, and the linear equation slope in step 11, find the distance of the screen L.

Pay attention to the units used in the experiment. It may be convenient to express the wavelength in mm.

14. The distance of the screen measured in the laboratory is 100 cm. Calculate the percent difference between this measured value and the value you determined from our diffraction experiment in Step 13.

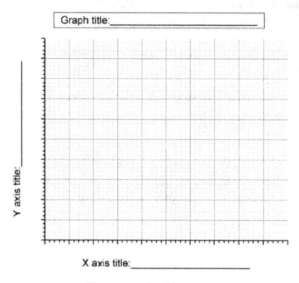

Figure 4.16. Graphing page.

4.3.3 Result and conclusion

Write a brief overview of what has been accomplished and concluded in this activity.

4.4 Real lab: *diffraction grating and circular aperture*

4.4.1 Objectives

The objectives of this lab are
- Study diffraction of light by a grating and a circular aperture.
- Apply the diffraction equation and determine the wavelength of a red laser.
- Measure the diameter of a circular aperture using the Airy disk and the second dark ring.

4.4.2 Supplies

Optics bench, light source, laser pointer, 200 line/mm diffraction grating, diffraction plate, two component holders, ray table base, diffraction scale, viewing screen, and rubber bands (figure 4.17).

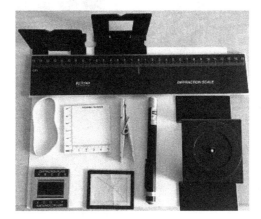

Figure 4.17. Supplies beside the optical bench.

4.4.3 Procedure

Most of the supplies listed are what we used in previous activities. Moreover, the first part is what we have already done in the virtual activity. The difference is the laser and the grating that we will use.

Part I: diffraction by a grating
The objective is to determine the wavelength of a red laser ($\lambda = 632$ nm) applying the diffraction equation for diffraction from a grating.
1. Affix the diffraction scale to one of the holders at the 0 cm mark on the optics bench. Make sure to center the 0 cm mark on the grating scale.

2. Affix the diffraction grating (with $n = 200$ lines/mm) to the second holder at the 10 cm mark on the optics bench.

3. Record the distance L between the grating and the diffraction scale with the uncertainties ΔL.

4. Turn the laser pointer and put it on the ray table base positioned on the optics bench about the 20 cm mark. We should see a diffraction pattern up to $m = 3$ on both sides of the 0 cm mark of the diffraction scale. The zero-order diffraction must overlap with the 0 cm mark (see figure 4.18).

Figure 4.18. Diffraction by a grating with $n = 200$ lines/mm.

5. Measure the distance Y, calculate $y = Y/2$, and record the corresponding values along with the uncertainties for the measured values (i.e. ΔY) in table 4.4. We do this up to the third-order ($m = 3$) (see figure 4.19).

Table 4.4. Measured and calculated values for diffraction from a grating.

m	Y	$y = \dfrac{Y}{2}$	ΔY	$\tan(\theta) = y/L$	$\theta = \tan^{-1}\left(\dfrac{y}{L}\right)$
1					
2					
3					

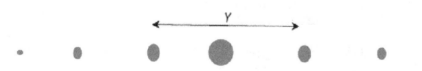

Figure 4.19. The schematic for the diffraction pattern up to $m = 3$.

6. Calculate the values for $\tan(\theta) = y/L$ and determine θ using the inverse tangent. Record the values in table 4.4.

7. Find the distance between adjacent lines for the grating.

8. Using the values for the third order of diffraction ($m = 3$), the distance between adjacent lines d, and the diffraction equation,

$$d \sin(\theta) = m\lambda, \qquad (4.17)$$

determines the laser's wavelength in the experiment.

9. Calculate the fractional uncertainty for L and Y (for $m = 3$) measurements. Using the highest fractional uncertainties, determine the uncertainty for the wavelength $\Delta\lambda$. Write the experimentally determined wavelength with the uncertainties.

10. The advertised value for the wavelength is $\lambda = 632$ nm. Does the experimental result for the wavelength in step 9 agree with the advertised value within the limits of uncertainties? If not, explain why!

Part II: diffraction by a circular aperture
In this part of the activity, we will apply what we have studied about the diffraction of light when it passes through a circular aperture. The goal of this activity is to determine the diameter of the circular aperture d, by measuring the radii of the Airy disk (r_1) and the first dark ring (r_2) for the diffraction of the same red laser ($\lambda = 632$ nm). We use the equations,

$$r_1 = \frac{1.22\lambda_n L}{d} \qquad (4.18)$$

and

$$r_2 = \frac{2.23\lambda_n L}{d}. \tag{4.19}$$

Hint: (a) You replace the grating with the diffraction plate (that has the circular aperture) and the diffraction scale by the viewing screen. (b) The recommended distance between the diffraction plate and the viewing screen is $L = 60$ cm to see a visible diffraction pattern. (c) When shining the laser through the diffraction plate, we see different diffraction patterns since there are different kinds of apertures. However, if we find the circular aperture, we see the diffraction pattern shown in figure 4.20.

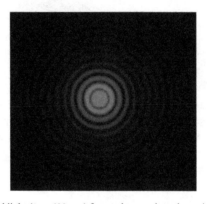

Figure 4.20. Diffraction of a red light ($\lambda = 632$ nm) from a laser pointer by a circular aperture with diameter d.

For this part of the lab, our job is to get this diffraction pattern on the viewing screen and answer the following questions.

1. Show the Airy disk and the second dark ring in figure 4.20 by drawing the corresponding circles.

2. On the viewing screen's diffraction pattern, trace the boundaries for the Airy disk and the second dark ring and use a ruler to measure the diameters for each (D_1 & D_2). Record the values with uncertainties of the measurements.

3. Calculate the fractional uncertainties FU_{D_1} and FU_{D_2}.

4. Determine the radii $r_1 = D_1/2$ and $r_2 = D_2/2$ with the uncertainties. Using these results, the viewing screen's distance $L = 60$ cm, and the relations in

equations (4.18) and (4.19), find the average diameter of the circular aperture d, with the uncertainties.

4.4.4 Result and conclusion

1. Is the value determined for the average diameter of the circular aperture acceptable? Explain why? Hint: Remember to see diffraction of light, the size of the aperture (i.e., d) must be comparable to the wavelength of the light (i.e., $\lambda = 632$ nm).

2. Write a brief overview of what has been accomplished and concluded in this real lab.

IOP Publishing

Virtual and Real Labs for Introductory Physics II
Optics, modern physics, and electromagnetism
Daniel Erenso

Chapter 5

Blackbody radiation

In the last chapters, we have studied the properties of light described using geometrical and wave optics. In this and the next two chapters, we will focus on the properties of light associated with its particle nature. Blackbody radiation, which is radiation emission by objects due to its temperature, is one example of such property of light. Wien's displacement law and the Stefan–Boltzmann law are two fundamental laws governing such radiation. In this chapter, after a brief introduction to these laws, we use the PhET simulation to better understand the blackbody radiation, in particular, Wien's displacement law. Wien's displacement law relates the object's temperature to the wavelength for the radiation's peak intensity. Using the PhTH simulation, we are interested in verifying the universal constant that relates these two quantities.

5.1 Basic theory

Up to this point, the properties of light we have studied involve conditions where ray optics and wave optics suffice to describe these properties of light. Next, we discuss other conditions where the particle nature of light is necessary.

Blackbody radiation
All objects emit radiation due to their temperature—the hotter the object, the more radiation energy they emit per unit of time. This emitted radiation consists of all wavelengths in the electromagnetic (EM) spectrum (not just the visible region!). The various wavelengths emitted by an object comprise what is called the emission spectrum of that object. When the object is in thermal equilibrium with its surroundings, it absorbs the same amount of energy that it is emitting. The various wavelengths of EM radiation emitted by a blackbody comprise the blackbody emission spectrum. Two examples of such a spectrum are shown in figure 5.1.

The two spectra in figure 5.1 correspond to two blackbodies having temperatures of $T_1 = 5790K$ (the Sun) and $T_2 = 3090K$ (incandescent bulb). The independent variable (horizontal axis) is the wavelength λ of the emitted radiation,

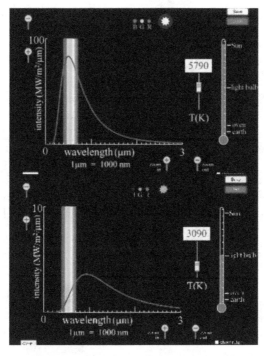

Figure 5.1. Blackbody spectrum from the Sun (top) and an incandescent bulb (bottom).

and the dependent variable is the intensity of radiation per unit wavelength interval, I_λ. (Remember that intensity is the energy per unit time per unit area: $J/(s \cdot m^2) = W/m^2$. Thus, I_λ must have MKS units of $(W/m^2)/m = W/m^3$). Note that the T_1 curve in figure 5.1 emits more radiation (per unit time per unit area) throughout the entire spectrum than the T_2 curve because of $T_1 > T_2$. Furthermore, each curve rises from zero, reaches a maximum, and then gradually decreases and eventually reaches zero again as λ approaches infinity. The higher-temperature ($T_1 = 5790K$) curve reaches a maximum at a shorter wavelength than the lower-temperature curve. The wavelength at which the blackbody spectrum reaches a maximum is denoted by λ_{max}. Max Planck found the equation which correctly describes the blackbody emission spectrum:

$$I(\lambda) = \frac{2\pi hc^2}{\lambda^5 \left(\exp\left(\dfrac{hc}{\lambda kT} \right) - 1 \right)}. \tag{5.1}$$

In equation (5.1), T is the temperature of the blackbody in Kelvin, c is the speed of light in a vacuum, $c = 3.0 \times 10^8 \, ms^{-1}$, k is Boltzmann's constant, $k = 1.38 \times 10^{-23} J/K$, and h is Planck's constant $h = 6.626 \times 10^{-34} J \, s$. Planck also found that radiant energy only came in little, discrete packets of energy, called photons. The energy in each packet depends on the wavelength of the radiation and is equal to

$$E = \frac{hc}{\lambda}. \tag{5.2}$$

Using $f = c/\lambda$ the energy of a photon in equation (5.2) can also be written in the form

$$E = hf. \tag{5.3}$$

Planck hypothesized that the radiant energy inside a blackbody comes in packets of energy $E = hf$ only when they interact with the blackbody walls. Albert Einstein extended this idea and proposed that all EM radiation comes in energy packets, which he called 'quanta of light' (later to be called photons) of energy $E = hf$. For this reason, the photon–energy equation (5.3) is called *the Einstein relation*.

Wien's displacement law
Wien's displacement law states that the value of λ_{max} changes (or is "displaced") as the temperature T (in Kelvin!) of the blackbody changes according to

$$\lambda_{max} T = \beta, \tag{5.4}$$

where $\beta = 0.002\ 9$ km is the Wien constant.

The Stefan–Boltzmann law
The Stefan–Boltzmann law states that the total intensity of radiation emitted by the surface of the blackbody at all wavelengths, I is related to the temperature by,

$$\text{Ideal blackbody emission, } I = \frac{E_{total}}{A \cdot t} = \sigma T^4, \tag{5.5}$$

where

$$\sigma = 5.67 \times 10^{-8} \text{ W m}^{-2} \text{ K}^{-4}. \tag{5.6}$$

Since real objects emit radiation less efficiently than a blackbody at the same temperature (since, by definition, a blackbody is an ideal emitter and absorber of radiation!), we introduce a (unitless) term called the emissivity e, the value of which ranges from near 0 to 1 for a perfect blackbody. Note that

$$I = \frac{E_{total}}{A \cdot t} = e\sigma T^4, \tag{5.7}$$

real blackbody emission.

5.2 Virtual lab: *blackbody radiation*

5.2.1 Introduction

The objectives of this virtual lab are to understand better Wien's law

$$\lambda_{\max} T = \beta, \tag{5.8}$$

and the Stefan–Boltzmann law

$$I = e\sigma T^4. \tag{5.9}$$

To this end, open the PhET simulation for this virtual lab by going to/clicking https://phet.colorado.edu/en/simulation/blackbody-spectrum. You will see a window for blackbody spectrum simulation lab shown in figure 5.2. Click play to open the simulation window shown in figure 5.3. This simulation window displays the

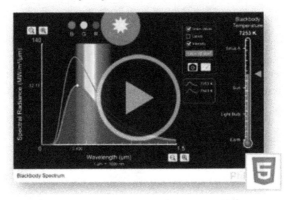

Figure 5.2. Blackbody spectrum simulation.

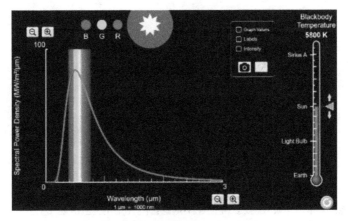

Figure 5.3. Blackbody spectrum.

intensity of the spectral power density (blackbody radiation energy per unit time, per unit area, and per unit wavelength) vs wavelength by different objects at different temperatures. The temperature can vary from about $300K$ (average Earth temperature) to about 11 000 K (a star hotter than our star, the Sun).

Note: do not click on any of the boxes or icons unless instructed to do so!

5.2.2 Part I: blackbody spectrum

1. Set the blackbody temperature to the Sun's surface temperature, about $T \simeq 5850K$, by dragging the temperature key. Check the 'graph values' and 'intensity' boxes. What are the values for the maximum spectral power density, the wavelength for the maximum spectral power density ($\lambda_{\max}$), and the maximum intensity (I). Establish a relationship between the area under the spectral power density curve and the intensity.

2. Is the wavelength for the maximum spectral power density ($\lambda_{\max}$) in the visible, infrared, or ultraviolet region of the electromagnetic spectrum?

   Check the box 'labels' and verify your answers.

3. Wien's displacement law states that $\lambda_{\max}$ is inversely proportional to the temperature of the blackbody,

$$\lambda_{\max} = \frac{\beta}{T}, \tag{5.10}$$

   where β is the Wien's displacement constant. Using the value for $\lambda_{\max}$ and the Sun's temperature, find the Wien's displacement constant. The result should be close to $\beta = 0.002\,9Km$.

4. Discuss what happens to $\lambda_{\max}$ and intensity when the blackbody temperature increases and decreases below the Sun's temperature. Change the temperature by dragging the temperature key up or down. A better view of the

spectral power density curve uses the zoom in and zoom out keys for the wavelength and the spectral power density axes.

5. Suppose we see two distant stars under a telescope, one is violet, and the other is red. Based on the observation in step 4, which star is hotter and why?

6. Select the light bulb temperature ($T \simeq 3050K$) and read the value for $\lambda_{\max}$. Is this wavelength in the visible, infrared, or ultraviolet region of the electromagnetic spectrum?

7. According to the Stefan–Boltzmann law the intensity of blackbody radiation is related to the temperature of the blackbody by

$$I = e\sigma T^4, \tag{5.11}$$

where

$$\sigma = 5.67 \times 10^{-8} \ \text{W m}^{-2} \ \text{K}^{-4}, \tag{5.12}$$

and e is the emissivity. For a perfect blackbody, $e = 1$. Assuming the bulb is a perfect blackbody, using the temperature $T \simeq 3050$ K for the light bulb, find this intensity. Does this value agree with the intensity read from the simulation? If not, what could be the reason for the difference between the two?

5.2.3 Part II: Wien's constant

In this part, we want to determine Wien's constant, $\beta = 0.002$ Km by doing a graphical analysis to a set of ($\lambda_{\max}$, T) data read from different blackbody emission spectral power density vs wavelength graphs. We collect 5–10 sets of data points for temperatures between $300K$ and $11\,000K$. Evenly distributed sets of data start at about a temperature $T \simeq 11\,000K = 11 \times 10^3 K$ and decrease with about an interval of $1500K$ or $2000K$ for a better graphical analysis. Zoom in or out both the spectral power density and wavelength axes to better view the graph. Make sure that graph values and intensity keys are on.

1. For the temperature values given in table 5.1, record the corresponding wavelength for the maximum spectral power density $\lambda_{\max}$ and the maximum value of the intensity I in the first and third columns.

Table 5.1. $\lambda_{\max}$ and I data for different temperature.

T (K)	$\lambda_{\max}$, (μm)	I ($\times 10^8$ W)	$\frac{1}{T}$, $\left(\frac{1}{K}\right)$	$T^4(K^4)$
11.0×10^3				
9.0×10^3				
7.0×10^3				
5.0×10^3				
3.0×10^3				
1.0×10^3				

2. Next, we will carry out a linear graphical analysis of the wavelength at which a blackbody emits the maximum intensity $\lambda_{\max}$, and the corresponding temperature T. According to Wien's Displacement law

$$\lambda_{\max} T = \beta, \tag{5.13}$$

where $\beta = 0.002\,9$ Km. Since the graph we make must be a linear graph, we note that

$$\lambda_{\max} = \beta \frac{1}{T} + 0. \tag{5.14}$$

3. Obviously, from equation (5.14), we see no direct linear relationship between $\lambda_{\max}$ and T. In order to get a linear graph of the form

$$y = mx + b. \tag{5.15}$$

Compare equations (5.14) and (5.15), identify x, y, m, and b in equation (5.14).

4. Using the temperatures in column one on table 5.1, calculate $1/T$, and record the values (using scientific notation) in the space provided in the fourth column.

5. Use Excel to make a linear graph and determine the equation for the best-fit line, and sketch the graph in figure 5.4.

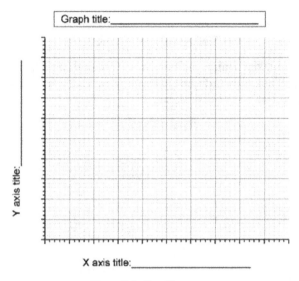

Figure 5.4. Graphing page.

6. The constant β in *Wien's displacement law* has a value $\beta = 0.002\,9$ Km, express this value in units of $K\,\mu$m.

7. What is the value of the constant β determined from the linear plot?

8. What is the percent difference between the actual value of β and the value determined in step 7?

5.2.4 Part III: Stefan–Boltzmann constant

1. According to the Stefan–Boltzmann law, the intensity of a blackbody radiation is related to the temperature by

$$I = e\sigma T^4, \tag{5.16}$$

where

$$\sigma = 5.67 \times 10^{-8} \text{ W m}^{-2} \text{ K}^{-4}. \tag{5.17}$$

Next, we shall determine the constant σ by making a linear graph analysis. For perfect blackbody radiation, we set the value for the emissivity $e = 1$. Under this condition, the intensity becomes

$$I = \sigma T^4 \tag{5.18}$$

or

$$y = mx + b, \tag{5.19}$$

where

$$y = I, \, m = \sigma, \, x = T^4, \, b = 0.$$

2. Find the value for $x = T^4$ using the temperatures in table 5.1 and record the values (using scientific notation).

3. Use Excel to graph y vs x (i.e., I vs T^4) and find the linear equation. Identify the slope and vertical intercept in the linear equation and determine the Stefan–Boltzmann constant σ.

4. Find the percent difference between the actual value of σ and the value you determined in step 3 (the experimental value).

5.2.5 Result and conclusion

Write a brief overview of what has been accomplished and concluded in this activity.

IOP Publishing

Virtual and Real Labs for Introductory Physics II
Optics, modern physics, and electromagnetism
Daniel Erenso

Chapter 6

The photoelectric effect

The last chapter introduced us to a particle of light (a photon). A photon carries energy $E = hf$, proportional to its frequency f, where $h = 6.626 \times 10^{-34}$ J s is Plank's constant. When we shine a monochromatic light on a polished metallic surface in a vacuum chamber, the photons can cause the emission of electrons depending on the photon's energy. This phenomenon is known as the photoelectric effect and the emitted electrons are referred to as photoelectrons. This chapter introduces us to the photoelectric effect. We begin with a brief introduction to the photoelectric effect followed by a PhET simulation of this effect in different metals, particularly sodium and zinc. This simulation examines the photoelectron's kinetic energy vs the photon incident on a sodium metallic plate. By carrying out a linear graphical analysis of these energies, we determine the cut-off frequency (the photon's minimum frequency required for sodium's photoelectric effect). We also study how this frequency varies from one metal to another by comparing it with zinc.

6.1 Basic theory

The photoelectric effect

Suppose a photon with frequency f and energy $E = hf$ is incident on a polished metallic surface (figure 6.1). This photon can cause the emission of an electron from the metallic surface if it carries enough energy. The emission of such electrons (photoelectrons) due to photons' absorption is known as the photoelectric effect.

The minimum energy of the photon required to emit an electron with zero kinetic energy is known as *the work function*, W, of the metal and is given by

$$W = hf_0 = \frac{hc}{\lambda_0}, \tag{6.1}$$

where f_0 is the cut-off frequency, $\lambda_0 = c/f_0$ is the corresponding cut-off wavelength, $h = 6.626 \times 10^{-34}$ J s is Planck's constant, and $c = 3.0 \times 10^8 \text{ms}^{-1}$, c is the speed of

doi:10.1088/978-0-7503-3715-1ch6

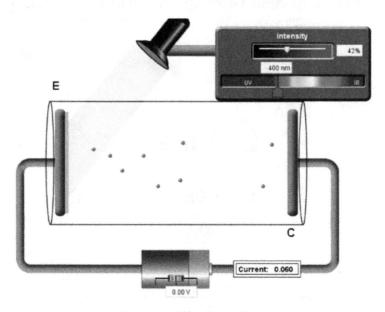

Figure 6.1. Photoelectric effect.

light in a vacuum. When $f < f_0$, there is no emission of electrons (or no photoelectric effect), and when $f > f_0$, the emitted electrons gain kinetic energy (KE) given by

$$KE = hf - W. \tag{6.2}$$

The electrons slow down and eventually stop when traveling in a region with an electrical potential gradient. The minimum potential difference (V) to stop a photoelectron, determined from conservation of energy

$$eV = KE \Rightarrow V = \frac{KE}{e}, \tag{6.3}$$

where e is the magnitude of the charge of an electron. Note that an electron volt (eV), $1eV = 1.6 \times 10^{-19}$ J, is a unit of energy.

6.2 Virtual lab: *photoelectric effect*

6.2.1 Introduction

The objectives of this virtual lab are

- To better understand the photon's energy, work function, cut-off wavelength (frequency), and kinetic energy of photoelectrons.
- To explore factors that cause a photoelectric effect in metals.
- To estimate the cut-off wavelength and frequency for sodium from the kinetic energy of the photoelectrons.

To this end, to open the PhET simulation for this virtual lab go to/click http:// phet.colorado.edu/en/simulation/legacy/photoelectric. You see a window for the photoelectric effect simulation lab shown in figure 6.2. Click play, and it downloads

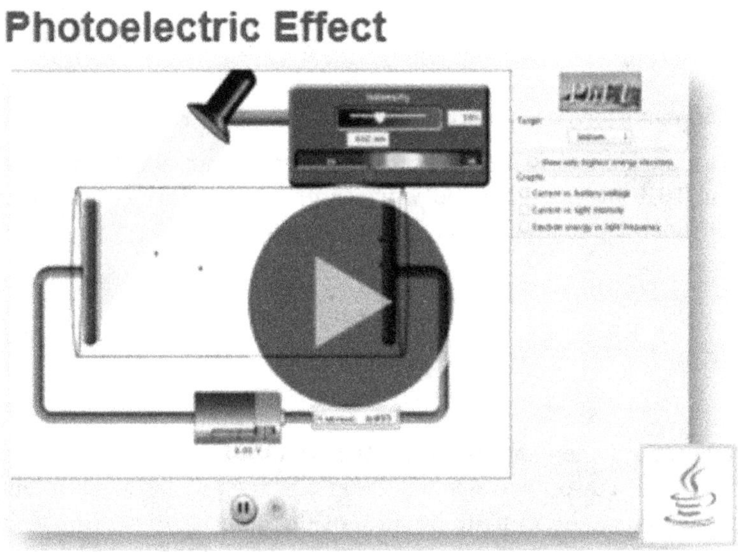

Figure 6.2. Photoelectric PHET simulation.

the simulation file. Open the file for the simulation window shown in figure 6.3. The photoelectric effect lab in figure 6.3 shows two separate parallel metallic plates placed in a vacuum chamber and connected to a battery by a conducting wire. Moving the key on the battery to the right or the left generates a current in the wire. On top of the simulation window, there is a light source that we can shine on one of the metallic plates. To control the wavelength and intensity of the light, slide the corresponding keys. The default setting of the target metal plate is sodium (top right corner).

Note: do not click on any of the boxes or icons unless instructed to do so.

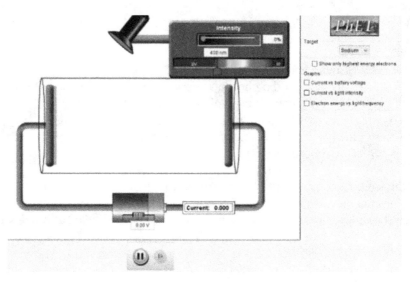

Figure 6.3. Photoelectric effect virtual lab.

6.2.2 Part I: photoelectric effect in sodium

1. Set the incident light wavelength to about 700 nm and then the intensity to about 10% by dragging the wavelength and intensity control keys. Does photoelectric effect take place? Explain why?

2. Move the wavelength key to the left (towards the smaller wavelength) and estimate the cut-off wavelength, λ_0, at which electrons begin to emerge from the metallic (sodium) plate because of the increase in energy of the incident photons. Estimate the uncertainties and calculate the fractional uncertainties for the wavelength.

3. Determine the cut-off frequency $f_0 = c/\lambda_0$, with the uncertainties. (Express the wavelength in meters.)

4. Using the cut-off frequency, calculate the work function,

$$W = hf_0, \tag{6.4}$$

where $h = 6.626 \times 10^{-34}$ J s is Planck's constant.

5. With the intensity still at 10%, set the incident photons wavelength to violet (400 nm) and watch the photoelectrons' speed ejected from the plate. Then change the wavelength to 100 nm and watch the speed of the photoelectrons again. Does the maximum kinetic energy of the photoelectrons increase or decrease as the wavelength of the incident photons decreases? Explain why?

6. Keep the wavelength 100 nm and the intensity at 10% and observe the emitted photoelectrons' number and speed carefully. Now change the intensity to 100% and observe what is happening. What is the effect of the change in intensity? Does the intensity affect the number or the kinetic energy of the individual photoelectrons? Give a brief explanation of why.

7. Set the wavelength back to 400 nm and the intensity back to 10%, change the target metal plate to zinc and wait for about 10 s. Do we see photoelectrons from the zinc plate? Give a brief explanation of why.

8. In order to see photoelectrons for the zinc plate, should we change the wavelength or the intensity? Should we increase or decrease this quantity? Give justification.

9. If we were to find the cut-off wavelength λ_0 and calculate the work function W for zinc, would it be greater or less than that of sodium? Give a brief explanation of why.

6.2.3 Part II: cut-off wavelength from photoelectrons KE

In this part, we want to determine the cut-off wavelength and frequency for the sodium target plate from the photoelectrons' kinetic energy. Reset the intensity to zero and the wavelength to 400 nm. Make sure the target metal is sodium. Click the box 'Electron energy vs light frequency' near the top right corner to open the window in figure 6.4.

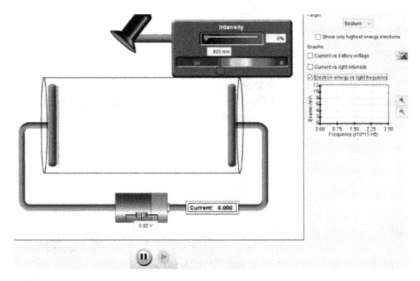

Figure 6.4. Photoelectric effect virtual lab with kinetic energy vs frequency graph.

We will set the incident photons' wavelength to the values listed in table 6.1 and read the photoelectrons' corresponding kinetic energy from the 'energy of the electrons vs light frequency' graph. The y-axis is the kinetic energy, and the x-axis is the frequency.

1. Set the intensity to 10% and the wavelength to the initial value (100 nm) and read the kinetic energy (in eV) for the electron emitted from the y coordinate of the blue dot in the graph. Record the value on the second column in table 6.1. Repeat this step for all wavelength values in table 6.1.

Table 6.1. Wavelength of the incident photons and the kinetic energy of the photoelectrons.

λ in (nm)	KE in (eV)	$\frac{1}{\lambda}$ in $\left(\frac{1}{nm}\right)$	$\frac{KE}{hc}$ in $\left(\frac{1}{nm}\right)$
100			
125			
150			
175			
200			
290			
375			

2. We now analyze the maximum kinetic energy (KE) of the photoelectrons vs the energy of the incident photons ($E_p = hc/\lambda$) graphically. The graph we make must be linear. To this end, we recall in the photoelectric effect

$$\text{KE} = E_p - W, \qquad (6.5)$$

where W is the work function, which is related to the cut-off wavelength λ_0 by

$$W = \frac{hc}{\lambda_0} \qquad (6.6)$$

and E_p is the energy of the incident photon with wavelength λ,

$$E_p = \frac{hc}{\lambda}. \qquad (6.7)$$

Using the equations for the work function and the energy of the incident photon, the maximum kinetic energy of the emitted electrons can be rewritten as

$$\text{KE} = \frac{hc}{\lambda} - \frac{hc}{\lambda_0}. \qquad (6.8)$$

Dividing this equation by hc,

$$\frac{\text{KE}}{hc} = \frac{1}{\lambda} - \frac{1}{\lambda_0}. \qquad (6.9)$$

We are interested in linear graphical analyses of the electrons' kinetic energy vs the wavelength of the incident photons. Obviously, from equation (6.9), we can see that there is no linear relationship between KE and λ. To get a linear graph of the form,

$$y = mx + b, \tag{6.10}$$

what vs what should we plot? That means what should be y, m, x, and b, comparing equation (6.10) to equation (6.9)?

3. We note that the value for hc in J m is

$$hc = 6.626 \times 10^{-34} \text{ J s} \times 3 \times 10^8 \text{ m s}^{-1} = 1.988 \times 10^{-25} \text{ J m}.$$

Using the relations

$$1 \text{ eV} = 1.6 \times 10^{-19} \text{ J}, \ 1 \text{ nm} = 10^{-9} \text{ m}$$

express hc in units of eV nm.

4. Using the values for the wavelengths in column 1 in table 6.1, calculate $1/\lambda$, and record the result in column 3.

5. Using the result in step 3 and the values for the kinetic energy in table 6.1 calculate KE/hc and record the results in column 4.

6. Make a linear graph and determine the equation for the best-fit line using Excel. Sketch the graph on the graph paper provided in figure 6.5.

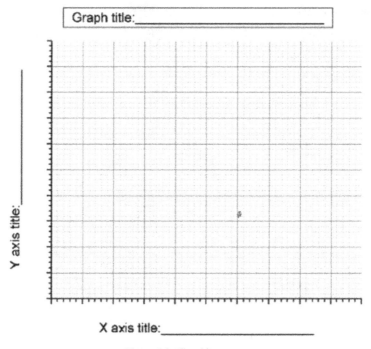

Figure 6.5. Graphing page.

7. Determine the cut-off wavelength value λ_0 by relating it to the slope or intercept in the linear equation determined in step 6. Using this wavelength, find the cut-off frequency f_0.

8. In steps 2 and 3 in Part I, we estimated the cut-off wavelength and frequency for sodium. Does the results for the cut-off wavelength and frequency determined in step 7 here agree with results in part I within the limits of uncertainties?

6.2.4 Result and conclusion

Write a brief overview of what we have accomplished and concluded in this activity.

IOP Publishing

Virtual and Real Labs for Introductory Physics II
Optics, modern physics, and electromagnetism
Daniel Erenso

Chapter 7

Introduction to atomic physics

Atomic physics is an area of physics where the immense application of quantum mechanics is required. The quantum mechanical description of an atom requires solving the Schrödinger equation. In this chapter, we are interested in the quantum description of the hydrogen atom. Our goal is not to solve the Schrödinger equation for the hydrogen atom but instead use the electron's quantum energy states, which results from solving the Schrödinger equation, and study the interaction of photons with atoms. We begin with a basic introduction to quantum mechanics, the Schrödinger equation's solutions for a hydrogen atom, and a quantum description of photons' interaction with the electron in hydrogenic atoms. We then use a PhTH simulation for a Bohr model to study a hydrogen atom's emission spectra resulting from interacting with white light. In the last part, we perform a real lab that uses a simple spectrometer to analyze the blackbody spectrum by an incandescent light bulb and atomic and molecular emission spectra of different atoms and molecules.

7.1 Basic theory

The hydrogen atom
The hydrogen atom with one proton and one electron makes up the smallest atom in the Universe. Its size is in the order of an angstrom ($A°$), 1 $A° = 10^{-10}$ m. In order to study the mechanics of the electron in a hydrogen atom, we need quantum mechanics. The equation that governs quantum mechanics is the Schrödinger equation:

$$O_{\text{KE}}\Psi(\vec{r},\ t) + O_{\text{PE}}\Psi(\vec{r},\ t) = O_{\text{TotalEnergy}}\Psi(\vec{r},\ t). \tag{7.1}$$

O_{KE} is the kinetic energy operator, O_{PE} is the potential energy operator, $O_{\text{TotalEnergy}}$ is the total energy operator, and $\Psi(\vec{r},\ t)$ is the wave function, which is a complex function in general.

doi:10.1088/978-0-7503-3715-1ch7

Time-independent Schrödinger equation
The time-independent Schrödinger equation obtained from the time-dependent Schrödinger equation above is

$$O_{\text{KE}}\Psi(\vec{r}) + O_{\text{PE}}\Psi(\vec{r}) = E\Psi(\vec{r}), \tag{7.2}$$

where E is the energy. For a hydrogen atom, the potential energy operator is

$$O_{\text{PE}} = -\frac{1}{4\pi\epsilon_0}\frac{e^2}{r}, \tag{7.3}$$

the kinetic energy operator is

$$O_{\text{KE}} = \frac{1}{r^2}\frac{\partial}{\partial r}\left(r^2\frac{\partial}{\partial r}\right) + \frac{1}{r^2\sin(\theta)}\frac{\partial}{\partial\theta}\left(\sin(\theta)\frac{\partial}{\partial\theta}\right) + \frac{1}{r^2\sin^2(\theta)}\frac{\partial^2}{\partial\varphi^2}, \tag{7.4}$$

and the wave function is separated into a radial component $R_{nl}(r)$ and angular component $A_{lm}(\theta, \varphi)$,

$$\Psi(\vec{r}) = R_{nl}(r)A_{lm_l}(\theta, \varphi). \tag{7.5}$$

Here r, θ, and φ are spherical coordinates describing the position of a point in space as shown in figure 7.1. In our case, the coordinates (r, θ, φ) represents the position of the electron relative to the position of the proton, which is at the origin $(r = 0, \theta = 0, \varphi = 0)$, and n, l, and m_l are quantum numbers. The quantum number n is the principal quantum number, l is the orbital angular momentum quantum number, and m is

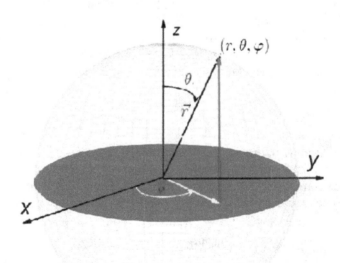

Figure 7.1. A point in space described by spherical coordinates r, θ, and φ.

the z component of the angular momentum quantum number. These quantum numbers take the following integral values:

$$\left\{\begin{array}{rl} n & = \ 1, 2, 3, 4, \ldots \\ l & = \ 0, 1, 2, 3, 4, \ldots(n - 1) \\ m_l & = \ -l, -(l - 1), \ldots 0, 1, 2, 3, \ldots(l - 1), l \end{array}\right\}. \tag{7.6}$$

In chemistry the n and l quantum numbers are described by

$$\left\{\begin{array}{rl} n & = \ 1, 2, 3, \ldots k, l, m, \ldots \\ l & = \ 0, 1, 2, \ldots s, p, d, f, \ldots \end{array}\right\}. \tag{7.7}$$

Solutions of the time-independent Schrödinger equation
We can solve the time-independent Schrödinger equation (which is a second order partial differential equation) using separation of variables and find the functions $R_{nl}(r)$ and $A_{lm}(\theta, \varphi)$ and the allowed energy states of the electron, E_n. We are not interested in solving the partial differential equation. Instead, we will focus on the physical interpretation of the wave function and the electron's energy in a hydrogen atom.

- *The radial function and the radial probability*: the probability of finding an electron in the state described by the quantum number n, l, m_l, at a distance r from the nucleus is given by

$$P_{nl}(r) = r^2 |R_{nl}(r)|^2. \tag{7.8}$$

- *The energy of the electron*: for an electron in the state described by the quantum numbers n, l, m_l the energy depends on the principal quantum number only. It is given by

$$E_n = -\frac{me^4}{8\epsilon_0^2 h^3}\frac{1}{n^2}, \tag{7.9}$$

$$E_n = -\frac{13.6 eV}{n^2} = -\frac{2.18 \times 10^{-18} J}{n^2}. \tag{7.10}$$

or

$$E_n = -\frac{hcR}{n^2}, \tag{7.11}$$

where $n = 1, 2, 3, \ldots$ and

$$R = 10\ 973\ 730\frac{1}{m} \tag{7.12}$$

is the Rydberg constant.

Photon emission and absorption

Consider an electron making a transition from an initial state i to a state f. Let the corresponding energies when the electron is in these states be E_i and E_f, respectively. Depending on which one of these energies is greater, photon emission or absorption of a photon will result (figures 7.2 and 7.3). The energy of this photon is related to these energy difference by

$$hf = \frac{hc}{\lambda} = E_f - E_i \quad \text{(absorption)}$$

$$hf = \frac{hc}{\lambda} = E_i - E_f \quad \text{(emission)}. \tag{7.13}$$

For a hydrogen atom, this becomes

$$hf = \frac{hc}{\lambda} = 13.6\text{eV}\left(\frac{1}{n_i^2} - \frac{1}{n_f^2}\right), \quad \text{absorption} \tag{7.14}$$

$$hf = \frac{hc}{\lambda} = 13.6\text{eV}\left(\frac{1}{n_f^2} - \frac{1}{n_i^2}\right), \quad \text{emission}. \tag{7.15}$$

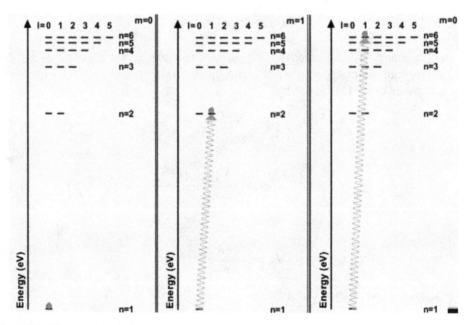

Figure 7.2. The energy levels for a hydrogen atom. From left to right: the electron is on the ground state ($n = 1$), first excited state ($n = 2$), and fifth excited state ($n = 6$). The transitions from the ground state to the excited state requires photon absorption.

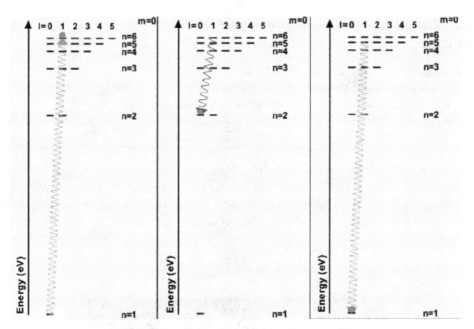

Figure 7.3. From left to right: the electron makes a transition from $n = 1$ to $n = 6$ (photon absorbed), from $n = 6$ to $n = 2$ (photon emitted) and from $n = 6$ to $n = 1$ (photon emitted).

Hydrogenic atoms

Hydrogenic atoms are ionized atoms in which all the electrons are removed from the atom except one. If the atomic number of these atoms is Z, then the energy of the electron is given by

$$E_n = -\frac{13.6 \text{ eV}}{n^2} Z^2. \tag{7.16}$$

7.2 Virtual lab: the *hydrogen atom*

7.2.1 Introduction

The objectives of this virtual lab are
- To understand the interaction of photons with an atom (a hydrogen atom).
- To study the energy of an electron in a hydrogen atom.
- To establish the relationship between photon's energy and electron's energy change during photons' absorption and emission.

   To this end, go to/click on https://phet.colorado.edu/en/simulation/legacy/hydrogen-atom to open the PhET simulation for this virtual lab and it opens a window for models of hydrogen atom simulation lab shown in figure 7.4. Click on the downward pointing arrow sign; it downloads the simulation file. Open the file to see the

Models of the Hydrogen Atom

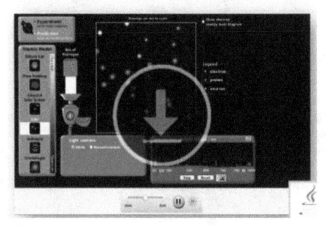

Figure 7.4. Atomic physics PHET simulation window.

simulation window shown in figure 7.5. The virtual lab window in figure 7.5 has three essential tools. One is the light source, which can produce a white light (multiple-wavelength) and a monochromatic wavelength (single wavelength), and you can turn it on and off. The second part is the hydrogen box right above the light source. The third is the 'experiment' and 'prediction' keys near the top left corner. Click on the 'Prediction' key and select 'Bohr'. That is the model we will be using throughout this virtual lab. Check the boxes 'show electron energy level diagram' and 'show spectrometer' boxes. At this stage, we will see what is shown in figure 7.6. Turn on the light source to produce multicolored photons (from a white light source) incident on the hydrogen atom. The spectrometer displays the type of photons (identified by their wavelength) emitted by the hydrogen atom due to the interaction with the incident photons. As long as the atom interacts with the incident photons, the number of photons emitted increases (forming a vertical stack of beads on the

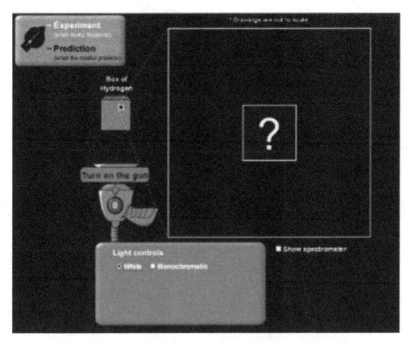

Figure 7.5. The simulation window for atomic physics virtual lab.

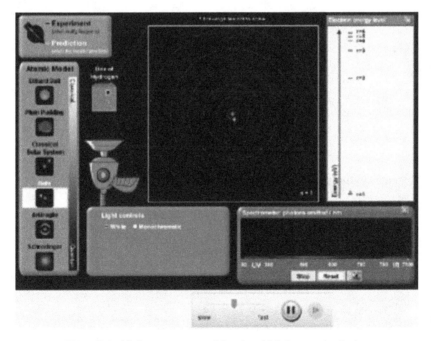

Figure 7.6. All the components of the virtual lab for atomic physics.

spectrometer at the photons' wavelength). Leave the spectrometer turned on for the whole duration of the lab activity. We can reset it whenever we need to. As time pass, we see four distinct lines are formed in the visible part of the spectra: purple, dark blue, cyan, and red. Ignore the UV and IR part of the spectra, as we cannot see it with our eyes in a laboratory setting. We can also take a snapshot of the spectra by clicking on the camera.

7.2.2 Part I: white light

1. Let the simulation run at slow speed and observe at the energy level diagram and also the spectrometer! What transition produces the emission of a red photon? Identify the initial n_i and the final energy level n_f for the electron that results in red photon emission. Find the corresponding energy values for each of these levels (E_{n_i} and E_{n_f}) and determine the energy difference between these two levels $\Delta E = E_{n_f} - E_{n_i}$. The energy of the electron at the nth energy level in eV can be determined using the relation

$$E_n = -\frac{13.6 \text{ eV}}{n^2}. \tag{7.17}$$

2. Convert the energy difference calculated in step 1 in eV into J.

3. Read the emitted red photon's wavelength from the spectrometer and calculate its energy. The energy of a photon E_p with wavelength λ is calculated using the relation

$$E_p = \frac{hc}{\lambda}, \tag{7.18}$$

where c is the speed of light in a vacuum

$$c = 3.0 \times 10^8 \text{ m s}^{-1} \tag{7.19}$$

and h is Planck's constant

$$h = 6.626 \times 10^{-34} \text{ J s}. \tag{7.20}$$

4. Reset the spectrometer but keep everything else the same. This time focus on the emission of blue photons. Identify the initial n_i and the final energy level n_f for the electron that results in a blue photon emission. Find the corresponding energy values for each of these levels and determine the energy difference between these two levels ΔE in eV. Convert the calculated energy difference into Joules.

5. Read the emitted blue photon's wavelength from the spectrometer and calculate its energy.

6. Compare the results for the energy difference ΔE when the electron makes the transition from the initial n_i to the final n_f levels, and the corresponding photons' emitted energy E_p and establish a relationship between ΔE and E_p. Take into account the uncertainties in reading the wavelength!
   (a) Express this relationship in terms of the wavelength of the photon, λ, and initial and final energies of the electron (E_{n_i} and E_{n_f})
   (b) Express this relationship in terms of the wavelength of the photon, λ, and initial and final principal quantum numbers for the electron (n_i and n_f).

7.2.3 Part II: monochromatic light

In Part I, we considered a white light source to excite the electron. A white light source has multiple wavelengths. In this part, we will consider a monochromatic light source when we have only one wavelength. To this end, stop the light source, click on monochromatic and reset the spectrometer.
   1. Make sure the prediction is still Bohr prediction, and no photons are heading to the hydrogen atom, and the electron is initially in the ground state $n_i = 1$.

2. The default setting for the monochromatic light source is 94 nm. Make sure that the wavelength is at this setting. Find the energy of the photon E_p with this wavelength in Joule and convert it into eV.

3. Turn the light source on, and as soon as the hydrogen atom absorbs a photon, turn it back off. In the process, observe what is happening to the electron's energy and the spectrometer. Repeat this process at least three times and describe the observation made.

4. Every time the hydrogen atom absorbs a photon with wavelength $\lambda = 94$ nm, it makes a transition from the ground level $n_i = 1$ to what level, n_f? Find the energy of these two levels in eV and determine the difference $\Delta E = E_{n_f} - E_{n_i}$. What is the relationship between the energy of the photon absorbed and the electron's energy difference ΔE.

5. After turning the light source off, we must have seen cases where the electron makes multiple transitions up to a maximum of five transitions in going back to the ground energy levels. Find the case where the electron makes three successive transitions in its way to the ground state, $n = 1$. For each of these transitions, identify the initial and final energy levels and calculate the corresponding energies. Record the values in table 7.1.

Table 7.1. Energies for the different energy levels in the electrons transition.

Transition	n_i	n_f	E_{n_i}	E_{n_f}	$|\Delta E| = \left|E_{n_f} - E_{n_i}\right|$
1st					
2nd					
3rd					
Total $|\Delta E|$					

6. Find the energy difference for the electron $|\Delta E|$ when it makes each of the three successive transitions and record the values in column six.

7. Calculate the total energy difference, Total $|\Delta E|$ and record the value in the space provided in table 7.1.

8. Use the results in step 7 to briefly discuss the relationship between the absorbed photon's energy E_p and the electron's energy change $|\Delta E|$.

9. Suppose we calculated the energy of the three photons emitted in the three successive transitions, E_{p_1}, E_{p_2} and E_{p_3}. Establish a relationship between these photons' energy and the electron's energy changes in the three successive transitions recorded in table 7.1, $|\Delta E|_1$, $|\Delta E|_2$, and $|\Delta E|_3$?

10. Establish a relationship between the total energy of the three emitted photons ($E_{p_1} E_{p_2}$ and E_{p_3}) and the energy of the absorbed photon E_p?

7.2.4 Result and conclusion

Write a brief overview of what we have accomplished and concluded in this activity.

7.3 Real lab: *spectroscopy*

7.3.1 Objective

The objectives of this lab are
- To see the application of diffraction grating in spectrometry.
- To measure the blackbody spectrum by incandescent light bulb (continuum spectrum).
- To observe atomic and molecular emission spectra (discrete spectra) of different atoms and molecules.

7.3.2 Supplies

The supplies you need for this activity (spectrometer, incandescent light bulb, and emission tubes) are shown in figure 7.7.

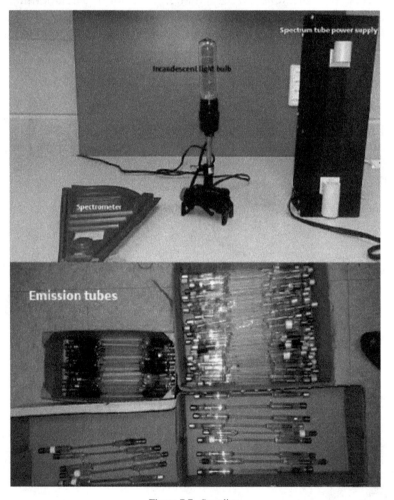

Figure 7.7. Supplies.

7.3.3 Procedure

This activity has two parts. In the first part we study the visible continuum spectrum of a blackbody radiation and in the second part we study the discrete spectra of different kinds of emission tubes made of different atoms or molecules.

7.3.4 Part I: the visible blackbody spectrum

You have studied the blackbody spectrum in chapter 5. In the simulation we used for blackbody radiation, from an incandescent light bulb ($T \simeq 3000$ K), the spectrum in the visible region is shown in figure 7.8.

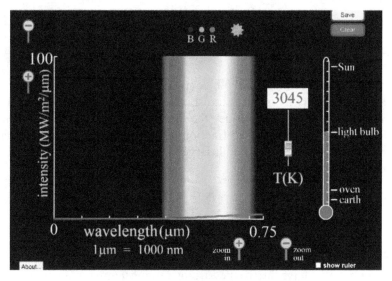

Figure 7.8. The visible spectrum of incandescent light bulb.

Our task is to determine the wavelength range for the seven different colors. We use a simple spectrometer made from a diffraction grating (project STAR spectrometer).

1. *Calibrating the spectrometer*: the strongest line visible in the emission spectrum for a fluorescent light bulb is *a bright green line*. This line corresponds to the $7s \rightarrow 6p$ transition in mercury with a wavelength of 546.1 nm (only one electron in mercury is excited and cascading downwards to a lower energy level in this transition). If the green line is not already lined up to the 546.1 nm mark, slide the color film in your spectrometer to the right or left to position this bright green line at 546.1 nm mark.

2. Focus the spectrometer window to the incandescent light bulb. You must see the rainbow colors shown in figure 7.8. Read the wavelength range of the seven different colors along with the uncertainties and write the values in table 7.2.

Table 7.2. Measured wavelength for incandescent light bulb emission spectrum.

Spectrum color	Wavelength range (nm)	Uncertainties (nm)
Red		
Orange		
Yellow		
Green		
Blue		
Violet		
Purple		

3. Explain why the colors in the blackbody spectrum are in the order that you observed them (why is the leftmost color on the left instead of on the right). (Hint: what equation describes the angle at which light at different wavelengths emerges from a diffraction grating (refer to the chapter about diffraction by a grating).) Will large wavelength light be deflected at large or small angles?

7.3.5 Part II: emission spectra

This part will study the emission spectra for at least two emission tubes of different atoms or gases. We have listed the strong atomic emission lines in the visible spectrum for argon (Ar), helium (He), hydrogen (H), krypton (Kr), mercury (Hg), neon (Ne), and xenon (Xe) atoms obtained from National Institutes of Standards and Technology (NIST). We also find the molecular emission lines for nitrogen (N_2) and oxygen (O_2) gases in this list.

1. Use one atomic emission tube (hydrogen or mercury are recommended) and one molecular emission tube (oxygen is recommended). Install these tubes (one tube at a time) into the emission tube power supply. Use the spectrometer to read the wavelength for at least five different color emission lines and record the values on the table provided.

2. Find the wavelength's actual values to each emission line measured for your emission tube from the list provided. This is the closest values to what you read with your spectrometer. Record the values in tables 7.3 or 7.4.

Table 7.3. Measured wavelength for tube 1 emission spectrum.

Emission tube one:		
Measured wavelength (nm)	Actual wavelength (nm)	% Difference

Table 7.4. Measured wavelength for tube 2 emission spectrum.

Emission tube two:		
Measured wavelength (nm)	True wavelength (nm)	% Difference

3. Calculating the percent difference between the actual and the measured values and record the results in tables 7.3 or 7.4.

4. For each atomic or molecular emission tube, suppose the percent difference for all wavelengths is nearly equal to zero. We can then make the conclusion that the atom or molecule inside the tube is actually what is labeled on the tube. Based on the results, are we comfortable to say the labeling on the tube is absolutely correct? Why?

5. We should have noticed when observing the discrete atomic emission spectra that we could also see a faint continuous blackbody spectrum in the background. Where do you think this blackbody spectrum is coming from?

7.3.6 Result and conclusion

Write a brief overview of what we have accomplished and concluded in this activity.

Hydrogen (H)
Wavelength (nm)
410.174
434.046 2
486.127 86
486.128 70
486.136 15
656.271 10
656.272 48
656.285 18
Mercury (Hg)
404.656 3
433.922 3
434.749 4
435.832 8
512.844 2
520.476 8
542.525 3
546.073 5
567.710 5
576.959 8
579.066 3
587.127 9
588.893 9
614.643 5

(*Continued*)

614.947 5
708.190
734.650 8
Oxygen (O_2)
440
490
525
540
550
565
615
660
665

Nitrogen (N_2)
Wavelength (nm)
410
420
425
440
445
500
505
520
530

540
550
560
580
585
590
600
615
620
625
630
640
650
660
670

Helium (He)
Wavelength (nm)
400.927
402.619 1
402.636
412.082
412.099

(Continued)

414.376
438.792 9
443.755
447.147 9
447.168
468.537 69
468.540 72
468.570 38
468.570 44
468.580 41
471.314 6
471.338
492.193 1
501.567 8
504.774
541.152
587.561 48
587.564 04
587.596 63
656.010
667.815 17
686.748
706.517 71

Argon (Ar)	
Wavelength (nm)	Wavelength (nm)
401.385 7	560.673 3
403.380 9	565.070 4
403.546 0	588.858 4
404.289 4	591.208 5
404.441 8	603.212 7
404.441 8	604.322 3
407.200 5	605.937 2
407.238 5	611.492 3
407.662	617.227 8
407.957 4	624.312 0
408.238 7	638.471 7
410.391 2	641.630 7
413.172 4	648.308 2
415.608 6	663.822 1
415.859 0	663.974 0
416.418 0	664.369 8
417.929 7	666.635 9
418.188 4	667.728 2
419.071 3	668.429 3
419.102 9	675.283 4
419.831 7	686.126 9
420.067 4	687.128 9
421.866 5	693.766 4

(*Continued*)

(Continued)

Argon (Ar)	
Wavelength (nm)	Wavelength (nm)
422.263 7	696.543 1
422.698 8	703.025 1
422.815 8	
423.722 0	
425.118 5	
425.936 2	
509.049 5	
514.178 3	
514.530 8	
516.577 3	
518.774 6	
521.681 4	
549.587 4	
555.870 2	

Neon (Ne)	
Wavelength (nm)	Wavelength (nm)
470.439 49	598.790 74
470.885 94	602.999 69
471.006 50	607.433 77
471.206 33	609.616 31
471.534 4	612.844 99
475.273 20	614.306 26
478.892 58	616.359 39

479.021 95	618.214 60
482.733 8	621.728 12
488.491 70	626.649 50
500.515 87	630.478 89
503.775 12	632.816 46
514.493 84	633.442 78
533.077 75	638.299 17
534.109 38	640.224 8
534.328 34	650.652 81
540.056 18	653.288 22
556.276 62	659.895 29
565.665 88	665.209 27
571.922 48	667.827 62
574.829 85	671.704 30
576.441 88	692.946 73
580.444 96	702.405 04
582.015 58	703.241 31
585.248 79	
587.282 75	
588.189 52	
590.246 23	
590.642 94	
594.483 42	
596.547 10	
597.462 73	
597.553 40	

Xenon (Xe)	
Wavelength (nm)	Wavelength (nm)
390.791	504.492
403.759	508.062
405.746	512.242
409.889	512.570
415.804	517.882
418.010	518.804
419.315	519.137
420.848	519.210
421.372	529.222
421.560	530.927
422.300	531.387
423.825	533.933
424.538	537.239
425.157	541.915
429.640	544.545
431.051	556.662
433.052	561.667
436.920	569.961
437.378	575.103
439.320	582.389
439.577	589.329
440.688	593.417
441.607	600.892
444.813	618.242

446.219	619.826
448.086	630.086
452.186	631.806
473.4152	634.396
479.2619	646.970
480.702	647.284
482.971	650.418
484.329	659.501
484.433	666.892
491.651	682.732
492.3152	688.216
497.171	694.211
497.271	697.618
498.877	699.088
499.117	708.215

IOP Publishing

Virtual and Real Labs for Introductory Physics II
Optics, modern physics, and electromagnetism
Daniel Erenso

Chapter 8

Introduction to nuclear physics

The last chapter introduced us to atomic physics, particularly to the electron's energy spectrum in hydrogen and hydrogenic atoms. This chapter introduces us to nuclear physics, in particular to radioactive decays. Radioactive decay is the transformation of a nucleus into another by emitting high-energy particles (such as alpha and beta) and high-energy photons (gamma). We begin with the fundamental theories to three types of radioactive decays (alpha, beta, and gamma decays) followed by a PhTH simulation to beta decay in hydrogen-3 and carbon-14 isotopes. We then do a real lab activity in cesium-137 isotope beta decay that produces an unstable barium-137 nucleus that eventually becomes stable by gamma decay. For these radioactive nuclei, we also determine the half-life time and decay constant.

8.1 Basic theory

Radioactive decay
Radioactive decay is the disintegration of a nucleus into another by emitting high-energy particles (such as alpha and beta) and high-energy photons (gamma).

Alpha decay
An alpha particle is simply a typical helium nucleus, $_2\text{He}^4$. It is sometimes written $_2\alpha^4$ or simply α. (α is the Greek letter 'alpha'.) An alpha particle, a combination of two protons and two neutrons, is stable and readily escapes from radioactive isotopes instead of other less-stable combinations of protons and neutrons. The transformation of an unstable radioisotopes nucleus resulting in alpha particle emission is known as alpha decay. Alpha decay of radium (Ra) results in radon (Rn) and an alpha particle

$$_{88}\text{Ra}^{226} \rightarrow _{86}\text{Rn}^{222} + _2 \alpha^4. \tag{8.1}$$

Generally, for an atomic nuclei Y with atomic number Z and mass number A that undergoes alpha decay, the decay equation is given by

$$_Z Y^A \rightarrow _{Z-2} X^{A-2} + _2\alpha^4. \tag{8.2}$$

Beta decay

A beta particle is just an electron and written in the form $_{-1}\beta^0$ or simply β^- or e^- (where the '−' denotes the charge; this is to differentiate from the anti-electron, or positron, which is commonly written in the form β^+ or e^+). (β is the Greek letter 'beta'.) In beta decay, an electron is emitted from the nucleus in a seemingly unlikely event since we do not usually think of electrons as residing in the nucleus! This apparent paradox is explained by the fact that a neutron itself undergoes beta decay:

$$_0 n^1 \longrightarrow _1 p^1 + _{-1}\beta^0 + _0 \bar{\nu}^0. \tag{8.3}$$

In the above equation, the last symbol is the Greek letter *nu* with a bar above it. It represents an anti-neutrino, apparently massless and chargeless particle that carries away energy and momentum. The electron can be emitted from the nucleus once a neutron in that nucleus undergoes beta decay. The proton produced in the beta decay of the neutron can subsequently absorb the electron in an inverse beta decay process called electron capture:

$$_1 p^1 + _{-1}\beta^0 \longrightarrow _0 n^1 + _0 \nu^0, \tag{8.4}$$

where the ν represents a neutrino. For an atomic nuclei Y with atomic number Z and mass number A that undergoes beta decay, the decay equation is given by

$$_Z Y^A \rightarrow _{Z+1} X^A + _{-1}\beta^0 + _0 \bar{\nu}^0. \tag{8.5}$$

Carbon-14 undergoes beta decay,

$$_6 C^{14} \rightarrow _7 N^{14} + _{-1}\beta^0 + _0 \bar{\nu}^0, \tag{8.6}$$

where C is carbon and N is nitrogen.

Gamma decay

A gamma particle (γ or γ^0) is a high-energy photon emitted by a nucleus. The energies of gamma rays can range up to several MeVs (mega electron-volts). We have already discussed the fact that quantum mechanics must govern the nuclei's behavior since they are at the atomic level or smaller. Nuclei consist of nucleons (protons and neutrons) confined to a small region of space. Whenever a particle is confined to a region of space, the quantum formalism results in distinct quantum states allowed for the particle with corresponding energies, all characterized by certain quantum numbers. This quantum effect is just as real for the nucleons confined inside the nucleus as electrons confined in the atom. However, since the protons in the nucleus are confined in very close proximity to the other protons in the nucleus, the protons' electrical force trying to push each other away is extremely large. Therefore, the nucleus' energy levels within the nucleus are correspondingly more immense than the electronic energies, and the difference in energy levels can be huge. Because of this, the photons (gamma rays) emitted by a nucleus as a nucleon transit from a higher energy level to a lower one can have enormous energies.

An isotope in an excited state is denoted by putting an asterisk by its chemical symbol. We would thus write, for example, C^{12*} to represent an excited carbon-12 nucleus.

Radioactive decay law
The number of isotopes $N(t)$ that undergoes radioactive decay is given by

$$N(t) = N_0 e^{-\lambda t}, \tag{8.7}$$

where λ is called the decay constant, and N_0 is the number of isotopes at the initial time $t = 0$.

Half-life time
Half-life time $(T_{1/2})$ is a time required for the number of isotopes to reduce by half. Suppose at $t = 0$, the number of nuclei is N_0. At a time equal to the half life time $(t = T_{1/2})$, the number of nuclei will be $N(t) = N_0/2$. Thus using the equation

$$N(t) = N_0 e^{-\lambda t}, \tag{8.8}$$

we may write

$$\frac{N_0}{2} = N_0 e^{-\lambda T_{1/2}} \Rightarrow \frac{1}{2} = e^{-\lambda T_{1/2}} \Rightarrow \ln \frac{1}{2} = -\lambda T_{1/2}, \tag{8.9}$$

which gives

$$T_{1/2} = \frac{\ln 2}{\lambda}. \tag{8.10}$$

Activity
Activity $(A(t))$ is the number of radioactive decays per second at a given time t; it can be determined using

$$A(t) = \lambda N(t). \tag{8.11}$$

8.2 Virtual lab: *beta decay*

8.2.1 Introduction

The objectives of this virtual lab are
- To get a better understanding of radioactive decay.
- To study hydrogen-3 and carbon-14 beta decay.
- To estimate the half-life time and decay constant for hydrogen-3.

To this end, go to/click on https://phet.colorado.edu/en/simulation/legacy/beta-decay to open the PhET simulation for this virtual lab shown in figure 8.1. Click on

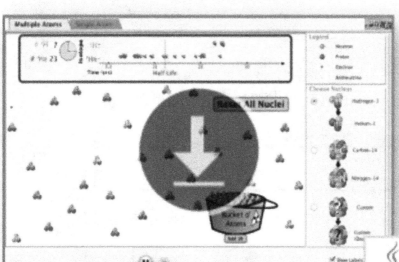

Figure 8.1. Beta decay simulation lab.

the downward pointing arrow to download the simulation file. Opening this file leads to the simulation window in figure 8.2 that simulates multiple nuclei beta decay, which we will study in Part II of this activity. For Part I, we are interested in single nucleus beta decay of hydrogen-3 and carbon-14 nuclei.

8.2.2 Part I: single atom beta decay

1. <u>Hydrogen-3:</u> click on the 'Single-atom' icon on the top left corner to begin the single hydrogen-3 nuclei decay process. Click on the 'Clear chart' icon on the top left corner. You will see hydrogen-3 nuclei in the center as shown in figure 8.3. Click on the 'Reset Nucleus' icon to see the beta decay process for hydrogen-3. The top left corner displays the decay time clock, which stops as

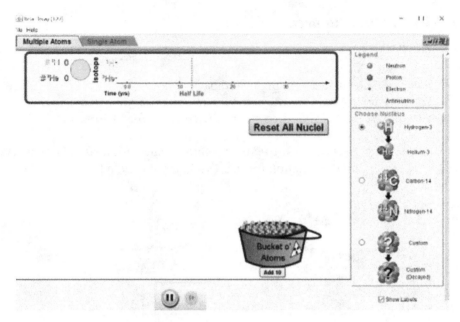

Figure 8.2. Beta decay simulation.

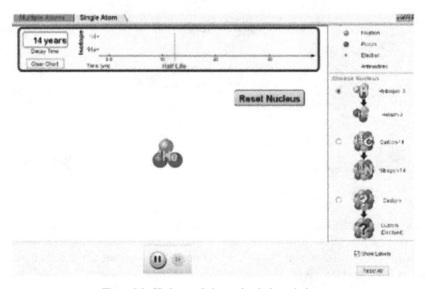

Figure 8.3. Hydrogen-3 decay simulation window.

soon as the nuclei undergo beta decay. Click this icon several times to understand the decay process better and quickly answer the questions below.

    (a) What are the number of nuclei (A), number of protons (Z), and number of neutrons (N) before the decay? (Use the color code in the legend on the top right corner.)

(b) We saw two particles emitted when the hydrogen nucleus decay. What are these particles? (You may want to pause the simulation by clicking the pause button.)

(c) What is the number of nuclei (A), the number of protons (Z), and the number of neutrons after the decay?

(d) Write down the decay equation. (You must use the correct symbol for the nuclei before and after the decay and the particles emitted.)

(e) Show that charge and mass are conserved in this decay process.

2. Carbon-14: click on 'Carbon-14' icon to study the decay process for single carbon-14 nuclei (see figure 8.4). Click on the "Reset Nucleus" icon to begin the beta decay process. The decay time clock stops as soon as the nuclei undergo beta decay. Here also click this icon several times to understand the decay process better.

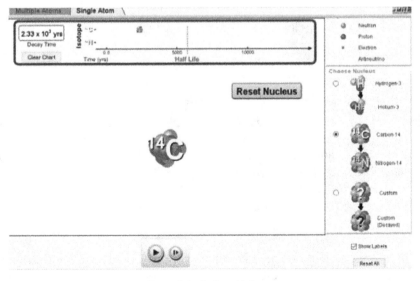

Figure 8.4. Carbon-14 decay.

(a) What is the number of nuclei (A), number of protons (Z), and number of neutrons (N) before the decay?

(b) What particles emitted when the nuclei decay?

(c) What is the number of nuclei (A), protons (Z), and neutrons after the decay?

(d) Write down the decay equation.

(e) Show that charge and the mass are conserved in this decay process.

(f) What is the half-life time for carbon-14 nucleus?

(g) Find the decay constant for carbon-14.

8.2.3 Part II: half-life time and decay constant

By now, from our observation in Part I, we should have noted that each nucleus that belongs to the same isotope does not decay at the same time. Suppose at the initial time the number of nuclei is $N_0 = 6.01 \times 10^{23}$ nuclei. How long should we wait for half of these nuclei to decay (i.e., $N(t) = N_0/2 = 3.005 \times 10^{23}$)? This time is what we call half-life time ($T_{1/2}$), and it is related to the decay-constant λ by

$$\lambda = \frac{\ln 2}{T_{1/2}}. \tag{8.12}$$

In this part of the activity, we are interested in estimating the half-life time and then find the decay constant for the hydrogen-3. To this end, we follow the steps listed below.

1. Click on the 'Multiple Atoms' icon and select hydrogen-3. Read the half-life time $T_{1/2}$ for hydrogen-3 from the time axis on the top part of the simulation window and calculate the decay constant.

2. We now estimate these values from a virtual experiment. To this end, click on the 'reset all' icon. Then click on the 'Add 10' icon to add ten hydrogen-3 nuclei into the bucket. These nuclei undergo beta decay at different times. Observe until all the nuclei have decayed, leaving behind ten daughter nuclei (^{3}He). Our goal is to find the average time for half of the nuclei to decay (the first five nuclei). This value is our experimental half-life time for $T_{1/2}$ hydrogen-3. To find this average half-life time: (a) Click on the 'Reset All Nuclei' icon. (b) Then observe the time axis until five hydrogen-3 nuclei undergo beta decay. At the instant the fifth nuclei decayed into ^{3}He, stop the simulation and read the time from the position of the fifth nuclei on the time axis and record this time in table 8.1, and (c) repeat steps (a) and (b) 20 times.

Table 8.1. Date table.

Trial	$T_{1/2}$	Trial	$T_{1/2}$	Trial	$T_{1/2}$	Trial	$T_{1/2}$
1		6		11		16	
2		7		12		17	
3		8		13		18	
4		9		14		19	
5		10		15		20	

3. Using the data in table 8.1 find the average half-life time $T_{1/2}$ for hydrogen-3 nuclei.

4. Find the experimental decay constant λ using the experimental half-life time.

5. Calculate the percent difference for the half-life time between what we read (step 1) and what we measured (step 4). Do the same thing for the decay constant. Based on the percent difference, what can we say about our experimental results?

8.2.4 Result and conclusion

Write a brief overview of what we have accomplished and concluded in this activity.

8.3 Real lab: *radioactivity and half-life time*

8.3.1 Objectives

The objectives of this lab are
(1) To determine the half-life of Ba-137 m.
(2) To understand radioactivity decay equations.
(3) To analyze data that varies exponentially with time.

8.3.2 Supplies

PASCO Cs-137/Ba-137 m isotope generator kit, the nucleus model 500 nuclear counter with Geiger tube and plastic tray (figure 8.5).

Figure 8.5. The Cs-137/Ba-137 m isotope generator kit, the nucleus 500 nuclear counter, and Geiger tube.

In this activity, we study radioactivity decay curves and the half-life of radioactive material.

Caution! This experiment uses weak radioactive sources; they are not dangerous. Special care must be taken not to spill the radioactive solution. Dispose of the sample used safely at the end of the lab. The Ba-137 will decay to the background in less than one hour.

8.3.3 Procedure

1. Cs-137 isotope beta decay produces the Ba-137 m. The 'm' in Ba-137m means that the newly formed barium atom's nucleus is in an excited state. Then the excited Ba-137 m nucleus emits energy via gamma decay and becomes stable. Write the two decay equations.

2. Measure the background count: set the high voltage to 400 V. With no sample in the detector, set the count interval to 5 min. Press COUNT to start

the background counting process, N_B. At the end of 5 min, record the background counts. Compute the background counts per minute ($n = N_B/5$ min).

3. The uncertainty in a counter reading (δN_B) is given by

$$\delta N_B = \sqrt{N_B}.$$

Compute the uncertainty.

4. The fractional uncertainty is given by

$$FU = \frac{\sqrt{N_B}}{N_B}.$$

Compute the fractional uncertainty. You will use it later on.

5. Set the timer to the 10-minute count interval. Place a sample containing Ba-137 m on the sample tray and insert it into the Geiger tube assembly, and when ready, press **START**. Record the readings every 1 min for 10 min until it decays to the background in table 8.2.

Table 8.2. Counts every 1 min.

Time (t) in minutes	Counts (N)
1.0	
2.0	
3.0	
4.0	
5.0	
6.0	
7.0	
8.0	
9.0	
10.0	

8.3.4 Data analyses

1. Put the time t and counts N in two columns (column A and column B) on an Excel sheet. Using Excel:

    (a) Calculate the counts per minute by finding the difference in each successive minute's counts and record the values in column C.

    (b) Subtract the background counts per minute from each of the values you recorded in column C and record the resulting values in column D.

    (c) Make a plot of the background corrected counts/min (i.e., the results in column D) vs time (column A).

2. The half-life time ($T_{1/2}$) is when the number of nuclei is reduced by half. Determine this time from the graph and write the value along with the uncertainty. To do this, we may follow the steps listed below.

    (a) Read the maximum number of counts per minute from your graph, N_{max} and the corresponding time, t_0.

    (b) Calculate half of the maximum value, $N_{max}/2$.

    (c) Locate the value for $N_{max}/2$ on your graph and read the corresponding time from the time axis, t_2.

    (d) Find the half-life time, $T_{1/2} = t_2 - t_0$.

3. Calculate the decay constant, λ, along with the uncertainty

$$\lambda = \frac{\ln 2}{T_{1/2}}.$$

4. Express the decay constant in SI units.

5. Sketch the graph on the graph paper provided in figure 8.6.

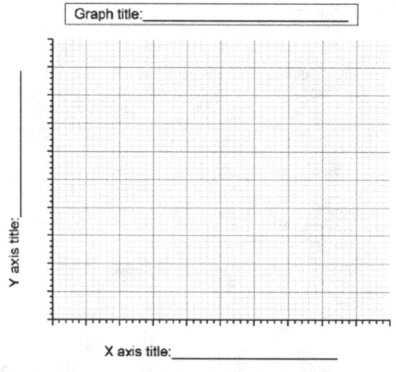

Graph title:_____

Y axis title:_____

X axis title:_____

Figure 8.6. Graphing page.

8.3.5 Result and conclusion

1. The actual value for the half-life time is 2.55 min. Does the experimental result agree with the actual value? If the answer is no, explain why?

2. Write a brief overview of what we have accomplished and concluded in this activity.

Virtual and Real Labs for Introductory Physics II
Optics, modern physics, and electromagnetism
Daniel Erenso

Chapter 9

Introduction to electronics

This chapter is the beginning of electromagnetism, which we will cover until the book's last chapter. We begin with introducing the fundamental theories such as Coulomb's law, electric field, electric potential, electric current, resistance, and properties of conductors and insulators. We will carry out two virtual labs using the PhTH simulation and one real lab. In the first virtual lab, we study Coulomb's law by examining how electrical force changes when we vary the charge and the distance between two objects. In the second virtual lab, we look into the properties of conductors and insulators of different items and use the measured electrical resistance to rank them from good to bad (insulators) conductor. This virtual lab will also introduce us to building a virtual electrical circuit, measuring the electrical potential difference between two points, and making a potential vs path graph in an electrical circuit. The real lab focuses on measuring various items' resistance (mostly what we considered in the virtual lab) using a multimeter. We also map the conducting path of a standard breadboard that we will be using to build a real electrical circuit in forthcoming activities.

9.1 Basic theory

Coulomb's law

Consider two stationary point charges q and Q, positioned in free space (vacuum) described by the position vectors, $\vec{r}'$ and $\vec{r}$, respectively (see figure 9.1). Coulomb's law states that there is an electrostatic force between the two charges that could be attractive or repulsive between these two charges. The magnitude of the force on any of these charges is directly proportional to the charges' magnitude and inversely proportional to the square of the separation distance. Vectorially, for example, the force on charge Q can be expressed as

$$\vec{F} = \frac{1}{4\pi\epsilon_0} \frac{qQ}{|\vec{r} - \vec{r}'|^2} \hat{R}, \qquad (9.1)$$

doi:10.1088/978-0-7503-3715-1ch9

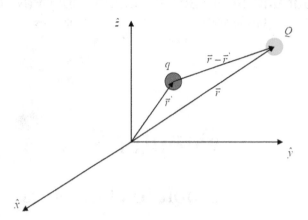

Figure 9.1. Two point charges in free space.

where

$$\epsilon_0 = 8.85 \times 10^{-12} \frac{C^2}{N\, m^2} \tag{9.2}$$

is the electrical permittivity of free space and $\hat{R}$ is the unit vector along the vector's $\vec{R} = \vec{r} - \vec{r}'$ direction. The magnitude of this force is expressible as

$$F = k\frac{|q||Q|}{R^2}, \tag{9.3}$$

where

$$k = \frac{1}{4\pi\epsilon_0} = 9.0 \times 10^9 \frac{N\, m^2}{C^2} \tag{9.4}$$

and $R = |\vec{r} - \vec{r}'|$ is the distance between the two charges q and Q.

The electric field
Consider an isolated object that we treat as a point particle. Suppose this particle is positioned at a point described by the position vector $\vec{r}'$ and carries a charge q. The electric field due to this charge at a position $\vec{r}$ is defined as the force per unit charge that would be exerted on a positive test charge Q when this test charge is placed at the position $\vec{r}$, and is given by

$$\vec{E} = \frac{\vec{F}}{Q} = k\frac{q}{R^2}\hat{R}, \tag{9.5}$$

where $R = |\vec{r} - \vec{r}'|$ is the distance between the charge q and test charge Q. The direction of the electric field depends on whether the charge is positive or negative. It is directed outward for a positive (figure 9.2 (a)) and inward for a negative (figure 9.2 (b)) charge.

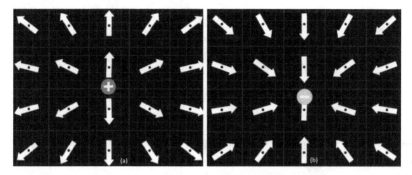

Figure 9.2. The electric field for a positive charge (a) and a negative charge (b). The white arrow with black dot shows the direction of motion for a positive test charge of unit magnitude.

The magnitude of the electrical force is related to the electric field by

$$F = |q_2| \, E = k\frac{|q_1||q_2|}{r^2}. \tag{9.6}$$

The electrostatic potential difference (voltage)
The electrostatic potential difference commonly referred to as voltage is the work done W_{if} per unit charge in moving a charge Q from a position i to a position f in space where there is an electric field

$$\Delta V = V_f - V_i = \frac{W_{if}}{Q}. \tag{9.7}$$

We measure voltage using a voltmeter or a multimeter in units of volts V,

$$1 \text{ Volt} = \frac{\text{Joule}}{\text{Coulumb}}. \tag{9.8}$$

Electrical grounding: a point the electric potential is zero, $V_G = 0$.
Conductors and insulators: for ideal conductors

$$\Delta V = V_f - V_i = \frac{W_{if}}{Q} = 0. \tag{9.9}$$

The electric current
An electric current is the rate of charge flow at a point in space or an electrical circuit. Suppose a total charge Q passes through a point in time t, the average electrical current is given by

$$I = \frac{Q}{t}. \tag{9.10}$$

It is measured using ammeter in units of ampere, A

$$1 \text{ Ampere} = \frac{\text{Coulomb}}{\text{sec}}. \qquad (9.11)$$

The conventional direction of current flow is the direction of positive charge flow.

Electrical resistance

In some materials, the electric current can flow easily almost with no resistance. Such materials are known as conductors. Other materials do not let the electric current pass through them at all. These materials are called insulators. The electrical resistivity of different materials is measured in terms of electrical resistance. The MKS unit of resistance is ohms that we denote with the Greek letter Ω. Other related common units used to measure resistance are $K\Omega$ and $M\Omega$

$$1 \text{ K}\Omega = 10^3 \text{ } \Omega, \ \ 1 \text{ M}\Omega = 10^6 \text{ } \Omega. \qquad (9.12)$$

9.2 Virtual lab I: *Coulomb's law*

9.2.1 Introduction

In our introduction to electronics, we have discussed Coulomb's law, which states that the magnitude of the electrical force (Coulomb's force F_C) between two charges q_1 and q_2 separated by a distance d is given by

$$F_C = k\frac{|q_1||q_2|}{d^2},$$

(9.13)

where

$$k = 9.0 \times 10^9 \frac{\text{N m}^2}{C^2}.$$

(9.14)

From this relation, we see that the force's magnitude has a direct relationship with the magnitude of the charges and an inverse square relationship with the distance between the two charges. This virtual lab aims to better understand Coulomb's law by examining these two relationships through graphical analyses. To this end, go to/ click on https://phet.colorado.edu/en/simulation/coulombs-law to open the PhET simulation window shown in figure 9.3.

Coulomb's Law

Figure 9.3. PhET simulation Coulomb's law.

Click play, and it will open another window, shown in figure 9.4(a), and click on Macroscale, and you will see the simulation window in figure 9.4(b). We see two spherical charged objects with charge q_1 and charge q_2. We can increase or decrease the charges of each using the right or left keys for each charge near the simulation window's bottom. There are two guys on the left and right sides holding these two charges to move them apart or close to change the two charges' distance.

Figure 9.4. Coulomb's law for macro and atomic scales simulation.

9.2.2 Part I: electrical force vs charge

1. Set the charge values for the two objects to $q_1 = -4\ \mu C$ and $q_2 = -4\ \mu C$.

2. Set the distance between the two charges to the smallest possible value, $d = 1.4$ cm, by moving the two guys closer. Line up the ruler's 0 cm mark to the center of the first charge and the 1.4 cm mark to the center of the second charge. Check the 'force value' and the 'Scientific notation' boxes. Read the magnitude and direction of the electrical force on q_2 by q_1 ($F_{q_2\ \text{by}\ q_1}$) and on q_1 by q_2 ($F_{q_1\ \text{by}\ q_2}$). Use '+' if the direction is to the right and ' −' if left. Pay attention to the direction and length of the two arrows that indicate the magnitude and direction of the electrical force vectors. Record the values in table 9.1.

Table 9.1. Electrical force vs charge data.

$q_2(\mu C)$	$F_{q_2\ \text{by}\ q_1}(\times 10^2\ N)$	Direction	$F_{q_1\ \text{by}\ q_2}(\times 10^2 N)$	Direction
−4.0				
−3.0				
−2.0				
−1.0				
0.0				
1.0				
2.0				
3.0				
4.0				

3. Keep the first object's charge $q_1 = -4\ \mu C$ and its distance from the second object, $d = 1.4$ cm. Increase q_2 with 1 μC, read the values for $F_{q_2\ by\ q_1}$ and $F_{q_1\ by\ q_2}$, according to the instruction in step 2. Record the values in the appropriate column in table 9.1.

   Note that 1 $\mu C = 10^{-6}$ C.

4. Examine carefully the data recorded in table 9.1 and answer the following questions!

   (a) What is the relationship between the magnitude of the electrical forces $F_{q_1\ by\ q_2}$ and $F_{q_2\ by\ q_1}$?

   (b) Based on the directions of the electrical forces, $F_{q_1\ by\ q_2}$ and $F_{q_2\ by\ q_1}$, what relationship can you establish about the nature of electrical force between the two charged objects: (i) when the charge of one object is negative, and the other is positive; (ii) when both charges are positive or negative; (iii) if one of the two objects is not charged at all (i.e., the charge is zero)?

5. Referring to equation (9.13) for the magnitude of the electrical force on q_1 by q_2 or on q_2 by q_1, we note that

$$F_C = F_{q_1\ by\ q_2} = F_{q_2\ by\ q_1} = k\frac{|q_1||q_2|}{d^2} = \left(\frac{k\,|q_1|}{d^2}\right)|q_2| = a\,|q_2| + b, \qquad (9.15)$$

where $b = 0$ and

$$a = \frac{k\,|q_1|}{d^2}. \qquad (9.16)$$

6. Using Excel and the data recorded in table 9.1, make a graph for the electrical force F_C vs $|q_2|$, determine the best-fit line equation, and sketch the graph in figure 9.5.

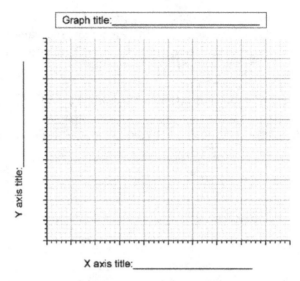

Figure 9.5. Graphing page.

7. We must have gotten a linear graph in step 6. What are the slope and vertical intercept of the linear graph?

8. *Finding the constant* k: by identifying the relevant results from step 7 (the slope or the intercept), whichever is related to k, find the equation for k (refer to equation (9.16)). Using this equation and the values for $|q_1|$ and d, find the value for the constant, k. Does the calculated value for k agree (or close enough) with the given value in equation (9.14)? If not, check the data in table 9.1.

9.2.3 Part I: electrical force vs distance

1. Set the charges for the two objects $q_1 = q_2 = 4 \, \mu$ C. Keep these values fixed.

2. Separate the two charges by a distance $d = 2.0$ cm, and read the electrical force's magnitude on q_1 by q_2 or on q_2 by q_1. (*You should know by now the magnitude of these two forces are equal.*) Record the value in the appropriate column in table 9.2.

Table 9.2. Electrical force vs distance data

Distance d (cm)	Coulomb's force F_C (N)	x
2.0		
4.0		
6.0		
8.0		
10.0		

3. By keeping the charges fixed ($q_1 = q_2 = 4 \mu$ C) and changing the distance with an increment of 2.0 cm, read the electrical force's magnitude between the two charges and record the values in the appropriate columns in table 9.2.

4. Using Excel, make a graph for F_C vs d for the data recorded in table 9.2. Based on what we see in the graph, discuss the relationship between F_C and d. Is this graph linear or nonlinear? Do we see an inverse relationship between the force and the distance? How does it differ from the force vs charge graph in Part I?

5. Referring to equation (9.13), when the two charges are positive and equal ($q_1 = q_2 = q$), we can rewrite the magnitude of the electrical force as

$$F_G = \frac{kq_1q_2}{d^2} = kq^2\frac{1}{d^2}. \tag{9.17}$$

In order to get a linear graph from the data we recorded in table 9.2, one must relate equation (9.17) to the general equation for a linear graph

$$y = ax + b. \tag{9.18}$$

Establish the relationship between equations (9.17) and (9.18) (i.e. $y = ?$, $a = ?$, $x = ?$, and $b = ?$) .

6. Calculate x using the correct relation determined in step 5, record the values in the appropriate column in table 9.2. Using Excel, make the linear graph using the appropriate values from table 9.2 and find the linear equation. Sketch the linear graph on the graphing page in figure 9.6.

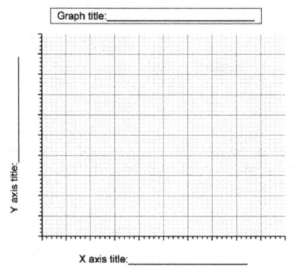

Figure 9.6. Graphing page.

7. What are the slope and the vertical intercept to the linear graph? It must include the appropriate units.

8. *Finding the charge*: using the relation we established for a, in step 5 and the slope for the linearized graph, find the charge $q_1 = q_2 = q$. Does the calculated value for q agrees (close enough) with the value you set for the charges (i.e., $q_1 = q_2 = 4\ \mu$ C)? If not, explain why not.

9.2.4 Result and conclusion

Write a brief overview of what we have accomplished and concluded in this activity.

9.3 Virtual lab II: *conductors and insulators*

9.3.1 Introduction

The objectives of this virtual lab are

- To study the electrical properties of different materials.
- To identify the common properties of materials, which are the very best conductors or the very worst conductors (insulators).
- To practice how to build virtual direct current (DC) circuits.
- To learn how to measure the potential differences between two points in an electrical DC circuit.

To this end, go to/click on https://phet.colorado.edu/en/simulation/circuit-construction-kit-dc to open the PhET simulation shown in figure 9.7(a), click play, and it will open another window shown in figure 9.7(b). Select 'Lab' and the simulation window for circuit construction kit will open up (see figure 9.8). We will use this circuit construction kit in the next upcoming DC circuits, and therefore, it is crucial to know all the elements of this kit. We see different electrical parts that we can use to build an electric circuit on the left side. It includes two batteries with different

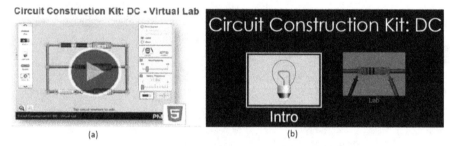

(a)   (b)

Figure 9.7. Circuit construction kit: DC—Virtual Lab

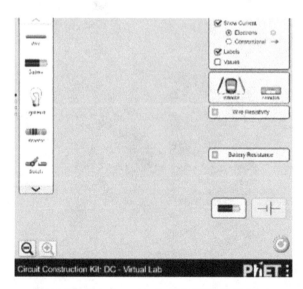

Figure 9.8. Circuit construction kit: DC—Virtual Lab

voltage, two light bulbs with different resistance, two resistors with different resistance, a switch, and a fuse. We also find different items with different resistance: a dollar bill, paper clip, coins, eraser, a hand, a dog, and a pencil (see figure 9.9). Use the up or down keys to see these electrical parts. To select any of these elements, we click on the element and drag it to the place we want it to be.

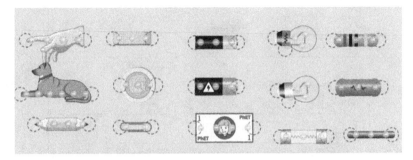

Figure 9.9. Accessible elements in DC circuit virtual labs.

9.3.2 Part I: conductors and insulators

1. Select all the elements listed in the first column on table 9.3.

2. On the top-right corner, click on 'Values', and you will see the measured resistance values in ohms (Ω). Record the values in the second column in table 9.3. In a real lab environment, we can measure the resistance of these materials by hooking a multimeter at the two ends of the material (marked by the dotted circles in figure 9.9)
   (Note: to set the switch turned on or off click on the switch).

3. Rank the materials listed in table 9.3 according to their electrical conductivity (from the very best to the very bad conductors). Record the ranks for each material in column 3 in table 9.3.

4. What are the very best conductors? What do these conductors have in common?

Table 9.3. Resistance data.

Item name	Resistance (Ω)	Rank
Light bulb-1		
Light bulb-2		
Switch turned on		
Switch turned off		
Eraser		
Hand		
Paper clip		
Coin		
Dollar bill		
Dog		
Pencil		
Resistor 1		
Resistor 2		

5. What are the very worst conductors (insulators)?

9.3.3 Part II: dc electrical circuit

In Part I, we have seen different materials and studied their electrical resistance, which we ranked from very best conductors to very bad conductors (insulators). In this part, we shall see how we build an electrical circuit using some of the elements and use a voltmeter to measure potential difference between two points in an electrical circuit.

1. Using one of the batteries, the switch, two of the resistors, and wires, build the circuit shown in figure 9.10. Make sure the switch is off at this stage.

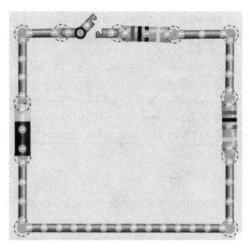

Figure 9.10. An electrical circuit with a battery, a switch, and two resistors. The wires used to build this circuit is copper wire with zero resistance.

2. Check the box 'value'; it will display the two resistors' resistance, including the switch and the potential difference across the battery (the source voltage). When we click on the resistors or the battery, it displays a small window near the bottom of the screen to change the resistance or source voltage values. Set the resistance for the two resistors to $R_1 = 33\Omega$ and $R_2 = 100\Omega$, ... Call the resistor between the horizontal wires in figure 9.10 R_1 and the other R_2. Set the source voltage $\Delta V_s = 1.5V$.

3. Turn the switch on and observe what happens to the electrons (the negative charges). Turn the switch off and observe what happens to the electron. Briefly discuss the observation made.

4. During the observation in step 3, we must have noticed that the boxes 'Show current' and 'electron' are checked on the screen's top-right corner. That means, under this condition, since it is the electrons that are moving and generate the electric current, we must have seen that the electrons are moving slowly in a counterclockwise direction. Check the box 'Conventional' and discuss what you see in the simulation.

Generally, it is essential to remember that when we study electrical circuits and talk about an electrical current, it always means the 'conventional' current.

Using voltmeter: we use a voltmeter to measure the potential difference between two points in a circuit. A voltmeter has a red lead (Red) and a black lead (Black). Consider two points A and B in an electrical circuit. Suppose we hook the red lead at point A and the black lead at point B. Then what we read from the voltmeter screen is the potential difference

$$\Delta V = V_{\text{Red}} - V_{\text{Black}} = V_A - V_B. \tag{9.19}$$

We usually represent the potential difference between two points as

$$\Delta V_{BA} = V_A - V_B. \tag{9.20}$$

A voltmeter reading could be positive or negative. If the reading for ΔV_{BA} is positive, point A is at a higher potential than point B, and if it is negative, point B is at a higher potential than point A.

   5. Keep the switch turned on and stay on conventional. Click on the battery symbol on the bottom right corner and you should see the circuit elements (the battery, resistors, and wires) replaced by the appropriate conventional symbols. The red arrows indicate the conventional current flow. You should see a circuit diagram shown in figure 9.11. We have labeled seven points named A, B, C, D, E, F, and G, in this figure.

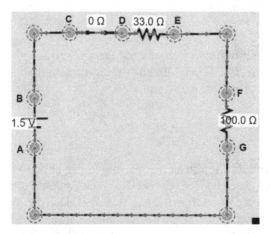

Figure 9.11. The circuit diagram.

   6. Click and drag the voltmeter close to the circuit we build to measure potential differences. To remove the voltmeter, we can click on and drag it back to its original place. The voltmeter is in the second box from the top on the screen's right-hand side.

7. Measure the potential differences between the pair of points listed in table 9.4. Refer to the diagram shown in figure 9.11 to locate the points. Find the total voltage

$$\Delta V_{\text{total}} = \Delta V_{AB} + \Delta V_{BC} + \Delta V_{CD} + \Delta V_{DE} + \Delta V_{EF} + \Delta V_{FG}, \qquad (9.21)$$

and record the values in table 9.4. We must find a value very close to zero!

Table 9.4. Measured potential difference.

Potential difference	Measured value in volts
$\Delta V_{AB} = V_B - V_A$	
$\Delta V_{BC} = V_C - V_B$	
$\Delta V_{CD} = V_D - V_C$	
$\Delta V_{DE} = V_E - V_D$	
$\Delta V_{EF} = V_F - V_E$	
$\Delta V_{FG} = V_G - V_F$	
Total voltage	

8. Using the measured values in table 9.4 make an electric potential vs path graph in figure 9.12.

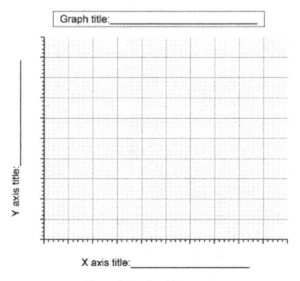

Figure 9.12. Graphing page.

9. *Without using the voltmeter* predict the reading for the potential differences listed in table 9.5. *Hint: refer to your measured values in table 9.4.*

Table 9.5. Predicted voltmeter reading.

Potential difference	Measured value in volts
$\Delta V_{BA} = V_A - V_B$	
$\Delta V_{CB} = V_B - V_C$	
$\Delta V_{DC} = V_C - V_D$	
$\Delta V_{ED} = V_D - V_E$	
$\Delta V_{FE} = V_E - V_F$	
$\Delta V_{GF} = V_F - V_G$	

9.3.4 Result and conclusion

Write a brief overview of what we have accomplished and concluded in this activity.

9.4 Real lab: *conductors and insulators*

9.4.1 Objectives

In this activity, we want to achieve the following objectives:
- Using a multimeter to measure resistance.
- Understand the difference between conductors and insulators.
- Map out the conducting paths in a circuit breadboard.

9.4.2 Supplies

The supplies we need for this activity are alligators, cables, a multimeter, goodies that include a dollar bill, paper clip, a coin, an eraser, a rubber band, a nail, Styrofoam, and pencil (for a complete list, see table 9.6). We also need a circuit breadboard shown in figure 9.13.

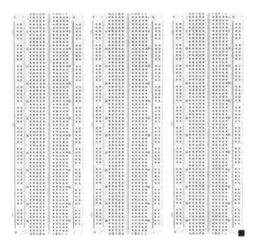

Figure 9.13. Circuit breadboard.

In figure 9.14 we see the different kinds of multimeters. We can use any one of these multimeters for measuring resistance or voltage for most of the real labs we will do.

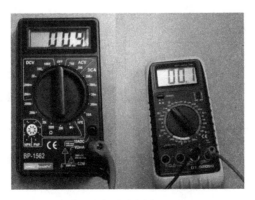

Figure 9.14. Multimeters.

9.4.3 Part I: using the multimeter to measure resistance

Measure the resistance for each of the items listed in table 9.6. Record the values (in units of Ω, $K\Omega$ ($=10^3\Omega$), or $M\Omega$ ($=10^6\Omega$)]) for each item in the second column in table 9.6. To measure resistance, connect the red and black cables to the multimeter

Table 9.6. Measured resistance for the different items.

Item name	Resistance
Air	
Thread	
Wire	
Toothpick	
Fingertip	
Paper	
Coin	
Styrofoam	
Paper	
Light bulb	
Aluminum foil	
Dollar bill	
Paper clip	
An eraser	
Resistor 1	
Resistor 2	
Resistor 3	
Fishing line	
Nail	
Rubber band	
Washer	

as shown and attach two alligators to the other ends of the cables and measure the resistance of the listed items (see figure 9.15).

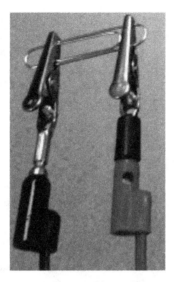

Figure 9.15. Measuring resistance.

9.4.4 Part II: mapping the breadboard

The circuit breadboard shown in figure 9.16 is where we build an electrical circuit. We must know the conducting and the non-conducting path on the breadboard. In this part of the activity, we will identify the conducting path.

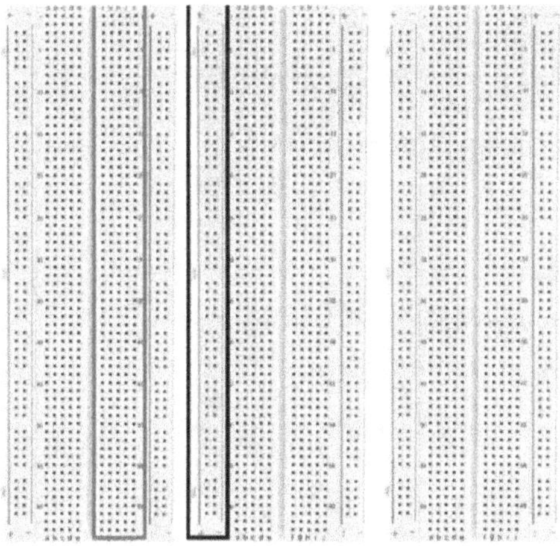

Figure 9.16. Mapping a circuit breadboard.

1. There are two different patterns of holes on the breadboard. These patterns are bounded by blue and black rectangular lines in figure 9.16. Our task is to determine the conducting path for the holes in the regions bounded by the blue and black lines.
2. Measure the resistance between holes using the multimeter. If the resistance is zero, the holes are connected by a conductor underneath, and there is a conducting path. If the resistance is infinity, no conductor is connecting the holes underneath, and therefore, there is no conducting path.
3. Show the conducting path for the holes in the blue and black rectangles by drawing a line connecting the holes if there is a conducting path in figure 9.16.

9.4.5 Part III: use the breadboard to build a circuit

Build the circuit shown in figure 9.17 on the breadboard. We will use the three resistors (R_1, R_2, and R_3) that we used in Part I and apply what we found as conducting and non-conducting paths of the breadboard in Part II.

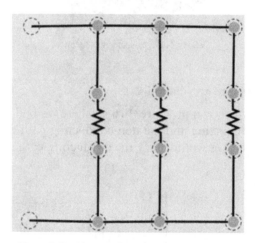

Figure 9.17. Three resistors connected in parallel.

9.4.6 Result and conclusion

1. What were the very best conductors that we investigated? What do these conductors have in common?

2. What were the very worst conductors (the best insulators)?

3. Write a brief overview of what we have accomplished and concluded in this activity.

IOP Publishing

Virtual and Real Labs for Introductory Physics II

Optics, modern physics, and electromagnetism

Daniel Erenso

Chapter 10

Resistors and Ohm's law

In the last chapter, we studied the electrical properties of different materials. Some materials, which we referred to as resistors, have finite, measurable resistance. We often see one or more resistors connected in different ways in electrical circuits. This chapter examines Ohm's law that relates the current through and the potential difference across a resistor. First, we give a brief introduction to Ohm's law and how we determine equivalent resistance for resistors connected in series or parallel. We then carry out a virtual lab to learn how to integrate an ammeter to measure the current through and a voltmeter to measure the potential difference (voltage) across a resistor in an electrical circuit. Using the measured current vs voltage and Ohm's law, we determine the circuit's resistance through a linear graphical analysis. We then replicate our virtual study in a real lab setting.

10.1 Basic theory

Ohm's law

Ohm's law relates the current I and the voltage ΔV_R in an ohmic device. It states that

$$\Delta V_R = IR, \tag{10.1}$$

R is called the resistance.

Series combination

Figure 10.1 shows two resistors connected in series. In a series connection, the equivalent resistance is given by

$$R_{12} = R_1 + R_2. \tag{10.2}$$

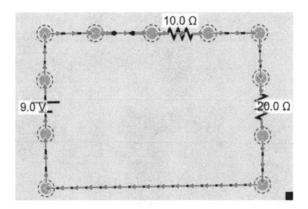

Figure 10.1. Resistors in series. The red arrows show the direction of the conventional current.

The equivalent resistance R_{12} replaces the two resistors, as shown in figure 10.2. In this particular case, the value for R_{12} is

$$R_{12} = 10 \ \Omega + 20 \ \Omega = 30 \ \Omega. \tag{10.3}$$

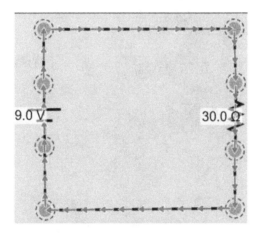

Figure 10.2. The equivalent resistor replaces the two resistors.

When resistors are in series, the same current flows through each resistor. Suppose the current through R_1 is I_1, through R_2 is I_2, and through the equivalent resistor R_{12}, is I_{12}, then

$$I_1 = I_2 = I_{12}. \tag{10.4}$$

Parallel combination
Resistors connected in parallel share the same nodes at both ends, as shown in figure 10.3. A node is a point in a circuit where two or more wires intersect.

The equivalent resistance R_{12} for two resistors R_1 and R_2 connected in parallel is given by

$$\frac{1}{R_{12}} = \frac{1}{R_1} + \frac{1}{R_2}. \tag{10.5}$$

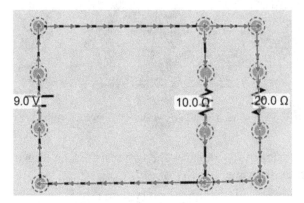

Figure 10.3. Resistors in parallel. The red arrows show the direction of the conventional current.

The equivalent resistance R_{12} replaces the two resistors, as shown in figure 10.4. In this particular case, the value for R_{12} is

$$\frac{1}{R_{12}} = \frac{1}{R_1} + \frac{1}{R_2} = \frac{1}{10\,\Omega} + \frac{1}{20\,\Omega} \Rightarrow R_{12} = 6.7\,\Omega \simeq 7\,\Omega. \tag{10.6}$$

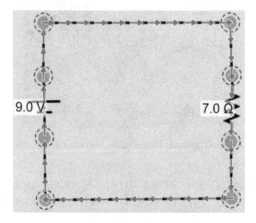

Figure 10.4. The equivalent circuit after the two resistors are replaced by the equivalent resistance.

When resistors are in parallel, the voltage across each resistor is the same. Suppose the voltage across R_1 is ΔV_1, across R_2 is ΔV_2, and across the equivalent resistor R_{12}, is ΔV_{12}, then

$$\Delta V_1 = \Delta V_2 = \Delta V_{12}. \tag{10.7}$$

10.2 Virtual lab: *Ohm's law*

10.2.1 Introduction

The objectives of this virtual lab are

- To learn how to connect an ammeter in a circuit to measure current in a resistor.
- To learn how to connect a voltmeter in a circuit to measure a potential difference across a resistor.
- To apply Ohm's law and determine the resistance of an ohmic device.

We use the same simulation kit that we used in the previous activity. Therefore, it is important to remember what we have introduced about this kit in the previous activity. To this end, go to/click on https://phet.colorado.edu/en/simulation/coulombs-law to open the simulation kit in figure 10.5(a). Click play and it will take you to the window shown in figure 10.5(b). Select Lab and you can open the window in figure 10.6.

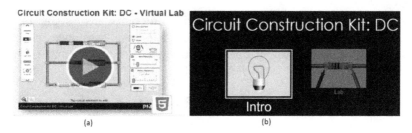

(a)   (b)

Figure 10.5. Circuit construction kit: DC—Virtual Lab.

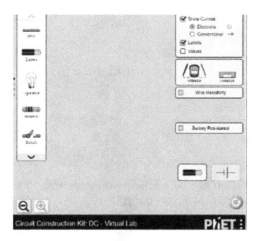

Figure 10.6. Circuit construction kit: DC—Virtual Lab.

10.2.2 Ohm's law

1. Using a battery, a switch, and the color-coded resistor, build the circuit shown in figure 10.7. Remember to select any of these elements, click on the element, and drag to the place we want to build the circuit.

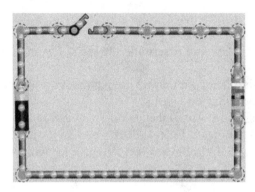

Figure 10.7. A battery, a switch, and a resistor connected by wires.

2. Set the source voltage (the potential difference across the battery) to $\Delta V_s = 0$. To change the source voltage or the resistance, click on the battery or the resistor. It opens a window where we can make changes.

3. Set the resistance of the resistor to $R = 47\ \Omega$.

4. How do we connect a voltmeter to measure the voltage across the resistor, ΔV_R?

5. How do we connect an ammeter to measure the current passing through the resistor, I?

6. Explains the difference between an ammeter and voltmeter in terms of how they are integrated into the circuit?

7. Connect the ammeter and the voltmeter to the circuit.

8. Now turn the switch on, check the boxes for 'Conventional', 'Values', and the battery symbol. For example, after performing this step, what we see on the screen is shown in figure 10.8.

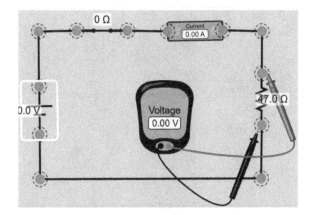

Figure 10.8. An ammeter and a voltmeter connected to the circuit.

9. Set the source voltage (the battery) to 0.5 V. Read the current I_R through and the potential difference (the voltage) ΔV_R across the resistor from the ammeter and the voltmeter, respectively. We must read non-zero values. Record the values in the appropriate columns in table 10.1.

Table 10.1. Measured current and voltage in the resistor.

Current I_R (A)	Voltage ΔV_R (V)

10. Increase the source voltage by another 0.5 V, read the current and the resistor's voltage until we reach 3.5 V.

11. Ohm's law relates the voltage across (ΔV_R), and the current (I) through a resistor with resistance, R, by the equation

$$I = \frac{\Delta V_R}{R}.$$
(10.8)

Using Excel, make a graph for I vs ΔV_R and determine the best-fit line equation. Sketch the graph on the graph paper provided in figure 10.9.

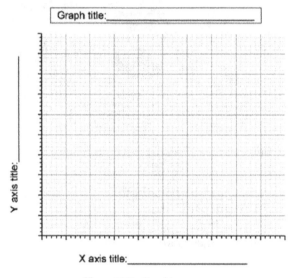

Graph title:_____

Y axis title:_____

X axis title:_____

Figure 10.9. Graphing page.

12. From the equation of the best-fit line, find the resistance of the resistor.

13. Does the result for the resistance determined from the graph agree with the value set initially ($R = 47\ \Omega$)? If not, we need to figure out why not. Find the percent difference.

10.2.3 Result and conclusion

Write a brief overview of what we have accomplished and concluded in this activity.

10.3 Real lab: *Ohm's law*

10.3.1 Objectives

The objectives of this lab are
   (a) To learn how to use a multimeter to measure dc voltage.
   (b) To learn how to use an ammeter to measure dc current.
   (c) To verify Ohm's law and study the properties of Ohmic devices.

10.3.2 Supplies

The supplies we need from the modular circuit basic kit included the 33 Ω and 100 Ω resistors, a battery, straight module, corner module, switch module, and jumper clips. We also need alligators, cables, a multimeter, and an ammeter (see figure 10.10).

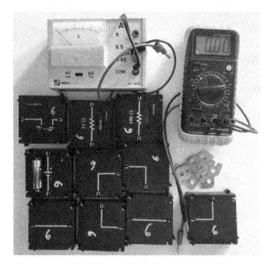

Figure 10.10. Supplies.

10.3.3 Part I: measuring resistance

   1. Measure the resistors' resistance in the resistor module ($R_1 = 33$ Ω and $R_2 = 100$ Ω) and record the measured values in table 10.2.

Table 10.2. Measured resistance.

	Resistance in Ω
Resistor R_1	
Resistor R_2	

2. Locate the ammeter and measure the resistance between the black and the red leads (figure 10.11) and record the values in table 10.3. There are three red leads ($0.05A$, $0.5A$, and $5A$).

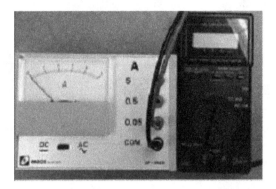

Figure 10.11. Measuring the resistance between the red and black leads.

Table 10.3. Measured resistance for the ammeter.

	Resistance in Ω
Red lead 1 and black lead	
Red lead 2 and black lead	
Red lead 3 and black lead	

10.3.4 Part II: measuring voltage

1. Using $R_1 = 33\ \Omega$, $R_2 = 100\ \Omega$, the battery ($\Delta V_s = 1.5\ V$), and the switch construct the circuit shown in figure 10.12. For additional help, see the constructed circuit shown in figure 10.13.
2. When we measure the voltage using a multimeter, what we read on the display screen is the potential difference between the red and the black leads. That means the voltmeter measures, $\Delta V_{BR} = V_R - V_B$. Turn the switch on and, using the multimeter, measure the potential differences (the voltage) between the next pair of points on the diagram shown in figure 10.12. Identify the corresponding closest points for A, B, C, D, E, F, and G on the circuit you constructed in figure 10.13. Record the values in table 10.4.

Find the total voltage

$$\Delta V_{total} = \Delta V_{AB} + \Delta V_{BC} + \Delta V_{CD} + \Delta V_{DE} + \Delta V_{EF} + \Delta V_{FG}$$

and record the values in table 10.4. You must find a value very close to zero!

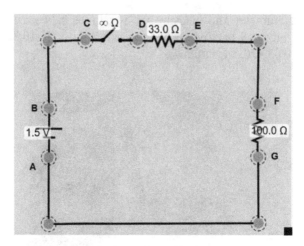

Figure 10.12. The circuit diagram.

Figure 10.13. The circuit with a battery, a switch, and two resistors.

Table 10.4. Measured potential difference.

Potential difference	Measured value in volts
$\Delta V_{AB} = V_B - V_A$	
$\Delta V_{BC} = V_C - V_B$	
$\Delta V_{CD} = V_D - V_C$	
$\Delta V_{DE} = V_E - V_D$	
$\Delta V_{EF} = V_F - V_E$	
$\Delta V_{FG} = V_G - V_F$	
Total voltage	

3. Using the measured values in table 10.4 make an electric potential vs path graph in figure 10.14.

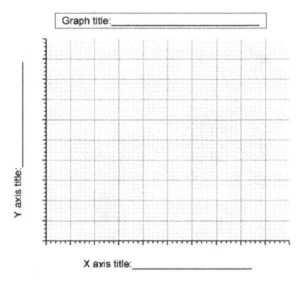

Figure 10.14. Graphing page.

10.3.5 Part III: measuring current

In this part of the lab, we are interested in verifying Ohm's law. According to Ohm's law, the current through a resistor is proportional to the resistor's voltage. We verify Ohm's law by measuring the current vs the voltage across the resistor and determine the resistance using graphical analyses. Therefore it is recommended to use a dc power source that we can vary the voltage output.

1. Construct the circuit shown in figure 10.15. The circuit shows a dc power source with a voltage ΔV_s, a switch, an ammeter, a resistor $R = 33\ \Omega$, and a

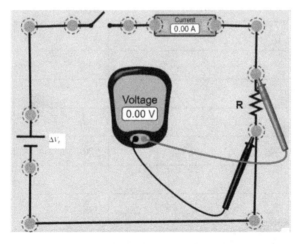

Figure 10.15. The circuit diagram.

multimeter. We can construct this circuit using the PASCO modules shown in figure 10.16. We can connect the red and black leads of the dc power source to the red and black leads on the left side of the circuit in figure 10.16.

Figure 10.16. The circuit constructed. The red and black leads on the left side are where we connect the dc power supply.

2. Turn on the power. Slowly increase the voltage across the resistor until we read 0.5 V on the multimeter screen. Read the current from the ammeter. We must read none zero value. Record the value in the first column in table 10.5.

Table 10.5. Current vs voltage data using dc power supply.

Current in Amps	Voltage in volts
0	0
	0.50
	1.50
	2.00
	2.50
	3.00
	3.50

3. By increasing the voltage to the next voltage values listed in the second column, read the corresponding currents from the ammeter, and record the values in the space provided on table 10.5.

Note: when we increase the voltage, at some point, the current will pass the 0.05A maximum value (see figure 10.17). In this case, we move to the 0.5 red lead.

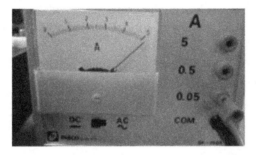

Figure 10.17. At higher voltage, the current will pass the 0.05A limit.

4. *Alternative way*: we can use the three batteries if we do not have a dc power supply, but we will be limited to three data points. We replace the module for power input with one battery, two batteries, and then three batteries modules and measure the voltage across and the current through the resistor (see figure 10.18) and record the values in table 10.6.

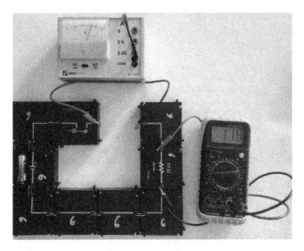

Figure 10.18. One battery used as a dc power source. We can add a second and then a third battery to increase the voltage across the resistor.

Table 10.6. Current vs voltage data using batteries.

Current in Amps	Voltage in volts
0	0

5. Estimate the uncertainties for the measurements.

6. Calculated the fractional uncertainties.

7. Ohm's law relates the voltage across (ΔV_R), and the current (I) through a resistor with resistance, R, by the equation

$$I = \frac{\Delta V_R}{R}.$$

(10.9)

8. Use Excel to make a graph for I vs ΔV_R and determine the best-fit line equation. Make a sketch of the graph on the graphing space in figure 10.19.

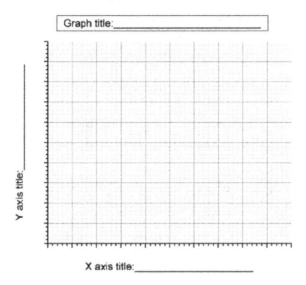

Figure 10.19. Graphing page.

9. From the equation of the best-fit line, find the resistance R along with the uncertainties.

10. Does the resistance determined from the graph agree with the resistor's measured value within the limits of uncertainties? If not, explain why?

10.3.6 Result and conclusion

Write a brief overview of what we have accomplished and concluded in this activity.

IOP Publishing

Virtual and Real Labs for Introductory Physics II
Optics, modern physics, and electromagnetism
Daniel Erenso

Chapter 11

Constant current circuit

In the previous chapter, we studied Ohm's law that relates the current through and the voltage across a resistor in an electrical circuit. Electrical circuits can have two or more resistors connected in different ways. This chapter studies electrical circuits with two or more resistors connected in series, parallel, or neither. We carry out two PhET simulations and two real labs. In the first simulation lab, we consider two resistors connected in series or parallel to a dc power supply. In the second virtual lab, we shall consider three resistors connected neither in series nor in parallel. In these virtual labs, we learn how to determine the current, the potential difference, and the power lost in each resistor. The real labs intended to replicate what we studied in the virtual labs. We will also introduce 'the backward' and 'Kirchhoff's' methods of predicting currents in an electrical circuit with multiple resistors. The current and voltage predicted using these methods are verified in a real experimental setting by measuring the current using an ammeter and the voltage using a multimeter.

11.1 Basic theory

In chapter 10, we were introduced to Ohm's law and determined the equivalent resistance when resistors connected in series or parallel. Here we focus on the methods to determine the current passing through, and voltage across, and power lost in resistors in an electrical circuit when more resistors are involved. To this end, it is essential to review the basic rules we were introduced to in chapter 10.

- *Ohm's law*: relates current (I) and voltage (ΔV_R) in an ohmic device

$$\Delta V_R = IR, \tag{11.1}$$

  R is called the resistance.
- *Series combination*: when two resistors are in series, the current is the same in each resistor,

$$I_1 = I_2 = I_{12}, \tag{11.2}$$

and the equivalent resistance is given by

$$R_{12} = R_1 + R_2, \tag{11.3}$$

- *Parallel combination:* when two resistors are connected in parallel, the electrical potential difference (voltage) across each resistor is the same,

$$\Delta V_1 = \Delta V_2 = \Delta V_{12}, \tag{11.4}$$

and the equivalent resistance is determined using the equation

$$\frac{1}{R_{12}} = \frac{1}{R_1} + \frac{1}{R_2}. \tag{11.5}$$

Whenever there are several resistors connected in series, parallel, or neither, to find the current through and then the potential difference across each resistor, we can follow the following two methods.

The backwards method
In this method, we find the equivalent resistance of the circuit, the source current, and work backward using the source current. While we work backward, apply the conditions
  (a) Current is the same for resistors connected in series.
  (b) Voltage is the same for resistors connected in parallel.
  (c) Ohm's law, $\Delta V_R = IR$.

The Kirchhoff's method
In this method, we find the current through each resistor by solving a set of linear equations obtained by applying Kirchhoff's voltage and current laws:
  (a) *Kirchhoff's voltage law:* start at one point in a circuit and go around any closed loop in the circuit. The voltage at the starting point of the loop must be the same as the voltage at the ending point since they are the same in the circuit. Another way of stating this is that the net change in voltage as we go around any closed loop in the circuit must be zero (so the starting potential must be the same as the ending potential). Kirchhoff's voltage law is just a restatement of the conservation of energy.
  (b) *Kirchhoff's current law:* the net current flowing into any node of a circuit must equal the net current flowing out of the same node. This law says that there is no place from which the charge magically appears in the circuit and no place that the charge magically disappears. Kirchhoff's current law is the conservation of charge in an electric circuit.

Power
Power supplied by a source or power lost in a resistor depends on the current I and the potential difference ΔV, and it is given by

$$P = \Delta VI \tag{11.6}$$

A resistor R in an electrical circuit loses energy in the form of heat. Applying Ohm's law, power lost in a resistor is also given by

$$P = \Delta VI = \frac{(\Delta V)^2}{R} = I^2 R. \qquad (11.7)$$

The energy lost in a resistor sometimes is referred to as Joule heat energy.

11.2 Virtual lab I: *resistors in series and parallel*

11.2.1 Introduction

The objectives of this virtual lab are
- To practice how to build an electrical circuit with two resistors connected in series or parallel.
- To learn how to find equivalent resistance for resistors connected in series or parallel.
- To get more experience in measuring current and voltage using an ammeter and a voltmeter in an electrical circuit with two resistors connected in series or parallel.
- To explore the relationship between power supplied by a battery and total power lost in resistors.

We will use the same circuit construction kit we used in the last two virtual labs. To open this kit go to/click on https://phet.colorado.edu/en/simulation/circuit-construction-kit-dc-virtual-lab and play the simulation (see figure 11.1).

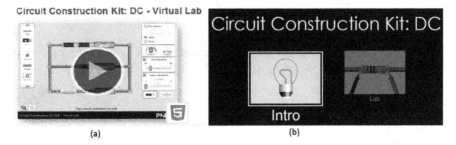

(a)   (b)

Figure 11.1. Circuit construction kit: DC—Virtual Lab

Select 'lab' and it will take you to the same window in figure 11.2.

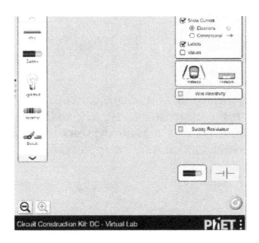

Figure 11.2. Circuit Construction Kit: DC—Virtual Lab

11.2.2 Part I: resistors in series

1. Using a battery, a switch, and the two resistors, build the circuit shown in figure 11.3. To select any of these elements, we need to click on the element and drag it to where we build the circuit.

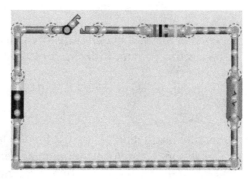

Figure 11.3. A battery, a switch, and two resistors connected by wires.

2. Set the source voltage $\Delta V_s = 5$ V, the resistance $R_1 = 47\ \Omega$, and $R_2 = 100\ \Omega$. We need to click on the battery or the resistor to change the source voltage or the resistance.

3. Are these resistors in series or parallel? Is it the current or the voltage that is the same for these two resistors? Find the equivalent resistance R_{12}.

4. Find the theoretical values for source current I_s, the current through, and the voltage across each resistor (I_1, I_2, ΔV_1, and ΔV_2). Record the values in the first column in table 11.1. Show the work in the space provided below. Record the values along with the appropriate units in table 11.1.

Table 11.1. Theoretical and experimental values.

	Theoretical value	Experimental values
I_s		
I_1		
I_2		
ΔV_s	5 V	5 V
ΔV_1		
ΔV_2		

5. Turn the switch on, check the boxes for 'Conventional', 'Values', and the battery symbol. After performing this step, we see what we see in figure 11.4.

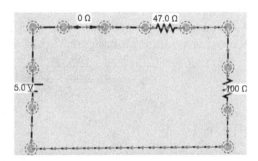

Figure 11.4. A battery, a switch, and two resistors connected in series.

6. Using the ammeter and the voltmeter, measure the current through (I_1 and I_2), the potential difference across (ΔV_1 and ΔV_2) each resistor (R_1 and R_2). What is the source current, I_s? Record the values in the second column in table 11.1. *Recall that an ammeter is connected in series while a voltmeter in parallel to measure current and voltage in an electrical circuit.*

7. Using the theoretical and the experimental results in table 11.1, find the theoretical and experimental power supplied by the source (P_s) and power

lost in the resistors (P_1, P_2). Show the work in the space provided below and record the values in table 11.2.

Table 11.2. Theoretical and experimental values.

	Theoretical value	Experimental values
P_s		
P_1		
P_2		

8. What is the total power lost in the resistors, and how does it relate to the power supplied by the source (battery)?

9. Use the measured values for the potential differences in table 11.1 to make an electric potential vs path graph in figure 11.5.

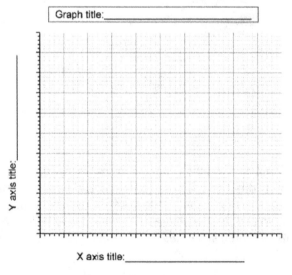

Figure 11.5. Graphing page.

10. Did all the measured currents and voltages agree with the theoretical calculations to within uncertainties? If not, identify which values did not agree with and explain what may cause the disagreement.

11.2.3 Part II: resistors in parallel

1. Without changing the source voltage $\Delta V_s = 5$ V and the resistance of the resistors $(R_1 = 47\ \Omega$. and $R_2 = 100\ \Omega)$ construct the circuit shown in figure 11.6.

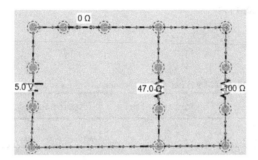

Figure 11.6. A battery, a switch, and two resistors connected in parallel.

2. Are these resistors in series or parallel? Is it the current or the voltage that is the same for these two resistors? Find the equivalent resistance R_{12}.

3. Find the theoretical values for the source current I_s, the current through, and the voltage across each resistor (I_1, I_2, ΔV_1, and ΔV_2). Show the work in the space provided below and record the values in table 11.3.

Table 11.3. Theoretical and experimental values.

	Theoretical value	Experimental values
I_s		
I_1		
I_2		
ΔV_s	5 V	5 V
ΔV_1		
ΔV_2		

4. Use the ammeter and voltmeter to measure the source current I_s, the current (I_1 and I_2), the potential difference (ΔV_1 and ΔV_2) for each resistor (R_1 and R_2).

Record the values in table 11.3. We should connect the ammeter and voltmeter at the right places in the circuit to measure the current and voltage.

5. Using the results in table 11.3, find the theoretical and experimental values for P_s, P_1, P_2. Show the work in the space provided below and record the values in table 11.4.

Table 11.4. Theoretical and experimental values for power.

	Theoretical value	Experimental values
P_s		
P_1		
P_2		

6. What is the total power lost in the resistors, and how does it relate to the power supplied by the source (battery)?

7. Use the measured values for the potential differences in table 11.3 to make an electric potential vs path graph in figure 11.7.

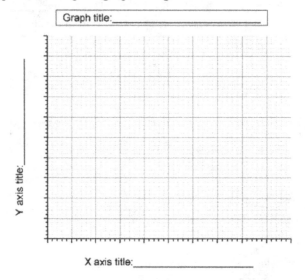

Figure 11.7. Graphing page.

8. Did all the measured currents and voltages agree with the theoretical calculations to within experimental uncertainties? If not, identify which values did not agree with and explain what may cause the disagreement.

11.2.4 Result and conclusion

Write a brief overview of what we have accomplished and concluded in this activity.

11.3 Virtual lab II: *mixed circuits and Kirchhoff's laws*

11.3.1 Introduction

The objectives of this virtual lab are

- To practice building electrical circuits with more than two resistors in series, parallel, and neither in series nor in parallel.
- To learn how to apply Kirchhoff's voltage and current laws to predict currents and voltage in an electrical circuit with nodes (junctions) and two or more closed loops.
- To further develop our understanding of power supplied by and lost in an electrical circuit.
- To have more experience in measuring current and voltage in a circuit with three or more resistors connected differently.

To this end, go to/click on https://phet.colorado.edu/en/simulation/circuit-con-struction-kit-dc-virtual-lab to open the circuit construction kit for this virtual lab. We use the same simulation we used in the last few virtual labs (see figure 11.8(a)). Click play and it will take us to the window shown in figure 11.8(b). Select 'Lab' to open the window shown in figure 11.9.

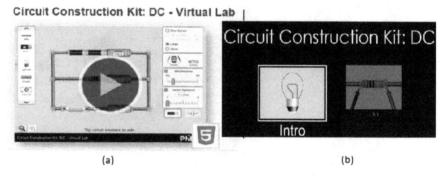

(a)   (b)

Figure 11.8. Circuit construction kit: DC—Virtual Lab

11.3.2 Mixed circuit

1. Using a battery, a switch, and four resistors $R_1 = 33\ \Omega$, $R_2 = 71\ \Omega$, $R_3 = 100\ \Omega$, and $R_4 = 33\ \Omega$ construct the circuit shown in figure 11.10. Set the source voltage to $\Delta V_s = 10$ V. *Note that $0\ \Omega$ is the resistance for the switch when it is on.*

2. Let the source current be I_s and the current through the resistors $R_1 = 33\ \Omega$, $R_2 = 71\ \Omega$, $R_3 = 100\ \Omega$, and $R_4 = 43\ \Omega$ be I_1, I_2, I_3, and I_4, respectively. Similarly, the voltage across each of these resistors be ΔV_1, ΔV_2, ΔV_3, and ΔV_4.

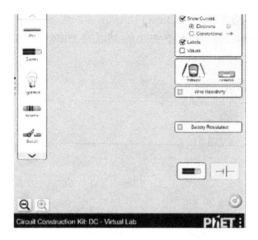

Figure 11.9. Circuit construction kit: DC—Virtual Lab

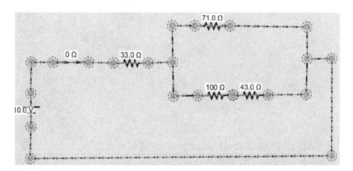

Figure 11.10. Mixed circuit.

(a) Identify the resistors connected in series or parallel. Is the current or the voltage for these resistors the same?

(b) Label the nodes (junctions) in figure 11.10 and write down the equation resulting from *Kirchhoff's current law*.

(c) Using two different closed loop in the diagram in figure 11.10 find two equations resulting from *Kirchhoff's voltage law.*

3. By solving the three equations, find the theoretical values for the current I_s, I_1, I_2, I_3, I_4, and voltage ΔV_1, ΔV_2, ΔV_3, ΔV_4. Show the work in the space provided below and record the values in table 11.5.

Table 11.5. Theoretical and experimental values.

	Theoretical value	Experimental values
I_s		
I_1		
I_2		
I_3		
I_4		
ΔV_s	10 V	10 V
ΔV_1		
ΔV_2		
ΔV_3		
ΔV_4		

4. Use the ammeter and voltmeter to measure the currents I_s, I_1, I_2, I_3, I_4, and voltages ΔV_1, ΔV_2, ΔV_3, ΔV_4. Record the values in the second column in table 11.5.

5. Using the theoretical and the experimental results in table 11.5, find the theoretical and experimental power supplied by the source (P_s) and the power lost in the resistors (P_1, P_2, P_3, P_4). Record the values in table 11.6.

Table 11.6. Theoretical and experimental values for power.

	Theoretical value	Experimental values
P_s		
P_1		
P_2		
P_3		
P_4		

6. What is the total power lost in the resistors, and how does it relate to the power supplied by the source (battery)?

7. Use the measured values for the potential differences in table 11.5 make an electric potential vs path graph in figure 11.11.

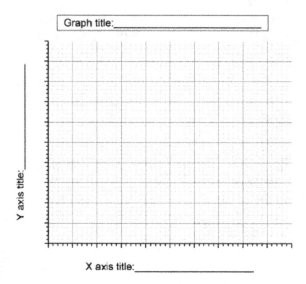

Figure 11.11. Graphing page.

8. Do all the measured currents and voltages agree with the theoretical calculations to within experimental uncertainties? If not, identify which values do not agree with and explain what may cause the disagreement.

11.3.3 Result and conclusion

Write a brief overview of what we have accomplished and concluded in this activity.

11.4 Real lab I: *resistors in series and parallel*

11.4.1 Objectives

The objectives of this lab are
- To study the properties of resistors connected in series or parallel.
- To learn how to use an ammeter to measure dc current in resistors connected in series or parallel.
- To learn how to use a multimeter (voltmeter) to measure dc voltage across resistors connected in series or parallel.

11.4.2 Supplies

The supplies we need from the modular circuit basic kit included the 33 Ω and 100 Ω resistors, the two batteries, straight module, tee module, corner module, switch module, and jumper clips. We also need alligators, cables, a multimeter, and an ammeter (see figure 11.12).

Figure 11.12. Supplies.

11.4.3 Part I: resistors in series

We will study the properties of two resistors connected in series (see figure 11.13).
  1. Measure the resistance for the two resistors ($R_1 = 33\ \Omega$, $R_2 = 100\ \Omega$). Record the measured resistance in the table below.

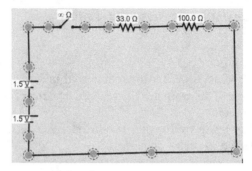

Figure 11.13. Two resistors and two batteries connected in series with a switch.

Resistor	Resistance in Ω
R_1	
R_2	

2. Measure the voltage across the two batteries and record the values below.

Battery	Voltage in Ω
1	
2	

3. For the circuit shown in figure 11.13 find the theoretical values for source current I_s, the current through and the voltage across each resistor (I_1, I_2, ΔV_1, and ΔV_2) when the switch on. Record the values with the appropriate units in the first column in table 11.7.

4. Now construct the circuit shown in figure 11.13. If you have difficulties you may refer to figure 11.14.

Table 11.7. Theoretical and experimental data.

	Theoretical value	Experimental values
I_s		
I_1		
I_2		
ΔV_s		
ΔV_1		
ΔV_2		

Figure 11.14. A constructed circuit with two batteries and two resistors connected in series. The circuit has a switch.

5. Turn the switch on and use the multimeter to measure the voltage across the batteries, $R_1 = 33\ \Omega$, and $R_2 = 100\ \Omega$ and show that the total voltage is zero. Make sure we move the leads either in a clockwise or counterclockwise direction for all (for example, see figure 11.15 for measuring the voltage across the two batteries).

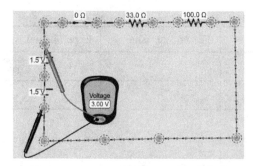

Figure 11.15. Measuring the voltage across the batteries using a multimeter (or a voltmeter).

6. Turn the switches off and integrate the ammeter (for example, see figure 11.16).

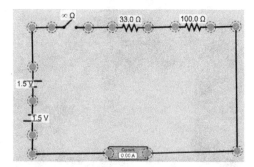

Figure 11.16. Resistors and an ammeter connected in series.

7. Turn the switch on, measure source current (I_s), the current through (I_1 and I_2) and the voltage across (ΔV_1 and ΔV_2) each resistor. Record the measured values in the second column in table 11.7.

8. What is the difference between connecting a voltmeter and connecting an ammeter in a circuit?

11.4.4 Part II: resistors in parallel

In this part of the lab, we study two resistors' properties connected in parallel as shown in figure 11.17.

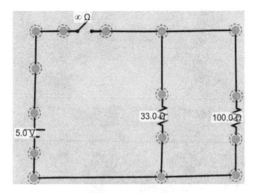

Figure 11.17. Resistors in parallel with two batteries and a switch.

1. For the circuit in figure 11.17, when the switch is on, find the theoretical values for the source current I_s, the current through and the voltage across each resistor (I_1, I_2, ΔV_1, ΔV_2). Record the values in the first column in table 11.8.

Table 11.8. Theoretical and experimental values resistors in parallel.

	Theoretical value	Experimental values
I_s		
I_1		
I_2		
ΔV_1		
ΔV_2		

2. Construct the circuit shown in figure 11.17. You can look at figure 11.18 if you run into difficulties.

3. Integrate ammeter and multimeter to the circuit, and measure the source current (I_s), the current through (I_1 and I_2) and voltage across (ΔV_1 and ΔV_2)

Figure 11.18. A constructed circuit with two batteries and two resistors connected in parallel. The circuit has a switch.

each resistor. Record the values in column 2 on table 11.8 (for example, the voltage and the current for the 33 Ω resistor measurement, see figure 11.19).

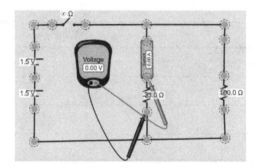

Figure 11.19. The integration of the voltmeter and the ammeter to measure the voltage and the current across the 33 Ω resistor.

4. Based on the results for the theoretical or experimental values establish the relation between I_1, I_2, and I_s (*Hint: Kirchhoff's current law*).

11.4.5 Result and conclusion

1. Did all the measured currents and voltages agree with the theoretical calculations to within experimental uncertainties? If not, identify which values did not agree and explain what may cause the disagreement.

2. Write a brief overview of what we have accomplished and concluded in this activity.

11.5 Real lab II: *mixed circuits and Kirchhoff's laws*

11.5.1 Objectives

The objectives of this lab are
- To practice building electrical circuits with more than two resistors neither in series nor in parallel.
- To learn how to apply Kirchhoff's voltage and current laws to predict currents and voltage in electrical circuits with nodes (junctions) and two or more closed loops.
- To develop a deeper understanding of power supplied and power lost in an electrical circuit.
- To have more experience in measuring current and voltage in mixed electrical circuit.

11.5.2 Supplies

The supplies we need from the modular circuit basic kit includes 33 Ω, 71 Ω, and 143 Ω resistors, two batteries (or a dc power supply), straight module, tee module, corner module, switch module, spring clip module, and jumper clips. We also need alligators, cables, a multimeter, and an ammeter (see figure 11.20).

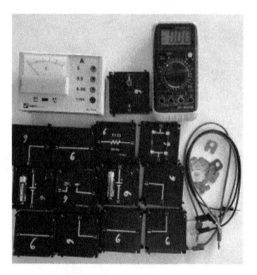

Figure 11.20. Supplies.

11.5.3 Mixed circuit

1. Measure the resistance for the three resistors ($R_1 = 33\,\Omega$, $R_2 = 71\,\Omega$, $R_3 = 143\,\Omega$). Record the measured resistance values in the table below.

Resistor	Resistance in Ω
R_1	
R_2	
R_3	

2. Using the multimeter, set the power supply voltage to 10 V.

3. Construct the circuit shown in figure 11.21. We can use a dc power supply that we control and set the source voltage $\Delta V_s = 10$ V; or you can use the two batteries with a total voltage $\Delta V_s = 3$ V. For example, using the two batteries in the circuit, we have constructed the mixed circuit shown in figure 11.22.

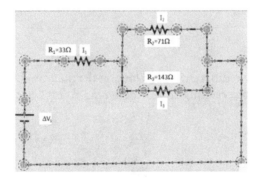

Figure 11.21. Mixed circuit design.

Figure 11.22. The circuit constructed.

4. Let the source current be I_s and the current through the resistors $R_1 = 33\ \Omega$, $R_2 = 71\ \Omega$, and $R_3 = 143\ \Omega$ be I_1, I_2, and I_3, respectively. Similarly, the voltage across each of these resistors be ΔV_1, ΔV_2, and ΔV_3.

    (a) Identify the resistors connected in series or parallel. Is the current or the voltage the same for these resistors?

    (b) Label the nodes (junctions) in figure 11.21 and write down the equation resulting from *Kirchhoff's current law*.

    (c) Using two different closed loop in the diagram in figure 11.10 find two equations resulting from *Kirchhoff's voltage law*.

5. By solving the three equations determined in steps 4 (b) and (c), find the theoretical values for source current I_s, the current through and the voltage across each resistor (I_1, I_2, I_3, and ΔV_1, ΔV_2, ΔV_3). Record the values in column one in table 11.9.

6. Integrate ammeter and multimeter in the circuit to measure the source current I_s, the current through I_1, I_2, and I_3, and the voltage across ΔV_1, ΔV_2, and ΔV_3 each resistor. Record the values in the second column in table 11.9.

Table 11.9. Theoretical and experimental values for current and voltage.

	Theoretical value	Experimental values
I_s		
I_1		
I_2		
I_3		
ΔV_1		
ΔV_2		
ΔV_3		

7. Using the theoretical and the experimental results in table 11.9 find power supplied by the source (P_s) and power lost in the resistors (P_1, P_2, P_3). Show the work in the space provide and record the values in table 11.10.

Table 11.10. Theoretical and experimental values for power.

	Theoretical value	Experimental values
P_s		
P_1		
P_2		
P_3		

8. What is the total power lost in the resistors, and how does it relate to the power supplied by the source (battery)?

9. Did all the measured currents and voltages agree with the theoretical calculations to within experimental uncertainties? If not, identify which values did not agree with and explain what may cause the disagreement.

11.5.4 Result and conclusion

Write a brief overview of what we have accomplished and concluded in this activity.

IOP Publishing

Virtual and Real Labs for Introductory Physics II
Optics, modern physics, and electromagnetism
Daniel Erenso

Chapter 12

Capacitor and RC circuits

The electrical circuit that we have studied up to this point involves only resistors. This chapter introduces dc electrical circuits with resistors and capacitors (RC circuits). We begin with a basic introduction to capacitors, electric current, and potential differences across a capacitor and a resistor in dc RC circuits during the charging and discharging of a capacitor. We then carry out a virtual lab using a PhTH simulation and a real lab replicating the virtual lab. In these labs, we study how a capacitor behaves for a short and long timescale. We also measure the potential difference across a capacitor as a function of time when charging and discharging in a dc RC circuit. Using these measurements, we determine the capacitive time constant using linear and nonlinear graphical analyses.

12.1 Basic theory

Capacitors
A capacitor consists of two conductors that are electrically isolated from one another, and it is useful to store energy. No current flows through a capacitor. The energy storing capacity of a capacitor depends on the capacitance C, given by

$$C = \frac{|Q|}{\Delta V_C},$$ (12.1)

where Q is the charge on and ΔV_C is the potential difference across the capacitor. Farads, F, is the SI unit for capacitance.

$$1\,\text{F} = \frac{\text{coulomb}}{\text{volt}} \Longrightarrow 1\,\text{F} = \frac{C}{V}.$$ (12.2)

Sometimes capacitance is expressed in units of

$$\mu\text{F} = 10^{-6}\,\text{F},\ \text{nF} = 10^{-9}\,\text{F},\ \text{pF} = 10^{-12}\,\text{F}.$$ (12.3)

A parallel plate capacitor: a parallel plate capacitor consists of two electrically isolated conducting plates. The capacitance depends on the plates' area, the separation distance, and the electrical permittivity of the material between the plates (see figure 12.1).

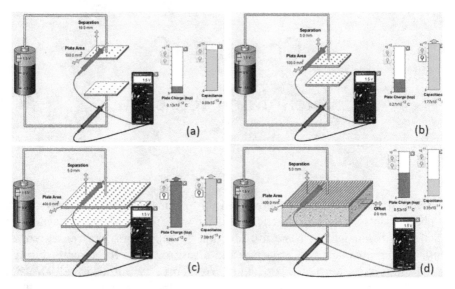

Figure 12.1. A parallel plate capacitor (a) with increased separation distance, (b) with decreased separation distance and increased plate area, (c) with decreased separation distance, and increased plate area, and (d) with decreased separation distance, increased plate area, and a dielectric inserted.

Series combination: two capacitors connected in series are shown in figure 12.2. The equivalent capacitance is given by

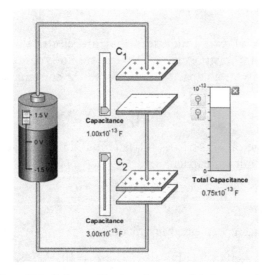

Figure 12.2. Series combination of two parallel plate capacitors.

$$\frac{1}{C_{12}} = \frac{1}{C_1} + \frac{1}{C_2}. \tag{12.4}$$

Parallel combination: two capacitors connected in parallel are shown in figure 12.3. The equivalent capacitance is given by

$$C_{12} = C_1 + C_2. \tag{12.5}$$

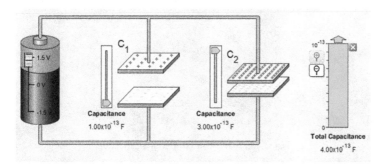

Figure 12.3. Parallel combination of two parallel plate capacitors.

RC circuit—charging capacitor:
Voltage: The potential difference (voltage) across a capacitor as a function of time is given by

$$\Delta V_C(t) = \Delta V_{max}(1 - e^{-t/\tau}), \tag{12.6}$$

where ΔV_{max} is the maximum voltage across the capacitor after a very long time, which will be just the voltage of the power source for the generic circuit of figure 12.4 (with the switch at position B):

$$\Delta V_{max} = \Delta V_s, \tag{12.7}$$

and the constant τ (the Greek letter tau) is called the *capacitive time constant*. This time constant has the units of seconds (s), and is given by

$$\tau = RC. \tag{12.8}$$

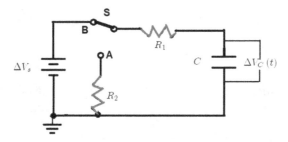

Figure 12.4. Dc RC circuit

(If there is only one capacitor in a circuit, but there is more than one resistor in the circuit, then the time constant of the circuit is equal to the equivalent resistance R_e, times the capacitance: $\tau = R_e C$.) A plot of $\Delta V_C(t)$ as a function of time is shown in figure 12.5 (the blue curve).

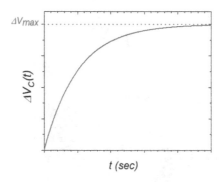

Figure 12.5. The voltage across a capacitor.

The current: to find the current in a charging capacitor we use Kirchhoff's voltage law for the circuit in figure 12.4 with the switch at **B**,

$$-I(t)R_1 - \Delta V_C(t) + \Delta V_S = 0 \tag{12.9}$$

and we get the following expression for the current I:

$$\frac{\Delta V_S - \Delta V_C(t)}{R_1} = I \tag{12.10}$$

RC circuits—discharging capacitor
Current: consider again figure 12.4. We assume the switch has been at position B for some time (so that there is a non-zero potential difference across the capacitor, as given by equation (12.6)), then throw the switch to position A. When we do this, we restart our stopwatch and start counting a new time, which we shall denote t'. (Thus, the switch is thrown to position A at the time $t' = 0$.) Using Kirchhoff's voltage law with the switch is now at position A instead of position B

$$\Delta V_R(t') + \Delta V_C(t') = 0 \tag{12.11}$$

from which we get, using Ohm's law $\Delta V_R(t') = I(t')R_2$,

$$I(t') = \frac{-\Delta V_C(t')}{R_2}. \tag{12.12}$$

Note that this equation shows us explicitly that the current in the circuit must be negative—that is, it flows counterclockwise around the circuit instead of clockwise. (Can you explain why the current is negative?) Again, from this equation, we can find the current in the discharging circuit if we know the voltage across the capacitor.

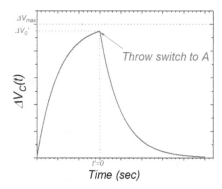

Figure 12.6. Charging and discharging voltage across a capacitor in dc RC circuit.

Voltage: the equation for the voltage across the capacitor, $\Delta V_C'$, as a function of time since the switch was thrown to position A (t'), for the discharging capacitor circuit is given by

$$\Delta V_C'(t') = \Delta V_o' e^{-t'/\tau'}. \tag{12.13}$$

In the equation above, the voltage $\Delta V_o'$ is the voltage across the capacitor when the switch is thrown from B to A (that is, at $t' = 0$), and τ' is the time constant in the discharging circuit. A graph for the function in equation (12.13) is shown in figure 12.6 (the pink curve A).

Half-life: the exponential growth or decay, denoted by the symbol $\tau_{1/2}$ is defined to be the amount of time for the growth to reach one-half of its limiting value (for example, the maximum voltage ΔV_{max} for the case of the voltage across the capacitor in a charging RC circuit), or for the decay to drop to one-half of its initial value.

12.2 Virtual lab: *capacitors and RC circuits*

12.2.1 Introduction

The objectives of this virtual lab are
- To build an RC circuit with switches and a dc power supply.
- To learn the properties of a capacitor when it is charging and discharging in an RC circuit.
- To better understand the capacitive time constant by experimentally analyzing the voltage vs time data.

To this end, go to/click on https://phet.colorado.edu/en/simulation/legacy/circuit-construction-kit-ac-virtual-lab to open the PhTH simulation window you see in figure 12.7. Download and open the file and it leads to the simulation window shown in figure 12.8.

Figure 12.7. Circuit construction kit (AC + DC), virtual lab shown in figure 12.8.

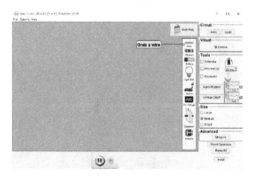

Figure 12.8. Opened circuit construction kit.

12.2.2 Part I: constructing an RC circuit with switches

1. Construct the circuit shown in figure 12.9. This diagram shows three resistors and one capacitor connected in series, two switches, and a battery. Set the

battery voltage to $\Delta V_s = 5$ V, the capacitance to the maximum value $C = 0.2F$ and each resistor's resistance to the maximum value $R_1 = R_2 = R_3 = 100\ \Omega$. Keep both switches open.

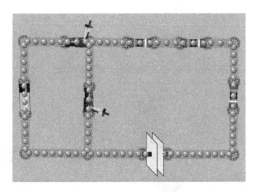

Figure 12.9. A direct current RC circuit.

For this kit, unlike the kit we used in the previous labs, to change the source voltage, the resistance, or the capacitance, we right click on the battery, the resistor, the capacitor, and a small window will be opened where we can select the changes we want to make.

2. Click on the box for a voltmeter and connect it across the capacitor to measure the voltage across the capacitor. We will measure the voltage across the capacitor when it is charging and discharging. (*Remember, for the voltmeter, we must connect the red lead to the positive side and the black lead to the capacitor's opposing sides*).

3. Click on the stopwatch box and reset it to zero to get ready for reading time and the voltage from the voltmeter. We will read every 15 s for 8 min. The first 4 min are for charging, and the next 4 min are for discharging the capacitor.

   Charging: when the first switch is closed and the second switch is open, the capacitor is charging (see figure 12.10(a)).

   Discharging: when the first switch is open and the second switch is closed, the capacitor is discharging (figure 12.10(b)). *Click on the switch's hand and drag it along the appropriate direction to open or close the switch.*

4. Now we are about to get started recording the voltage reading on the multimeter. For the first 4 min, keep the first switch closed and the second open. At $t = 4$ min $= 240$ s, open the first and close the second. During the entire 8 min, we read the voltage every 15 s and record the values in table 12.1.

   We recommend pausing the simulation at the end of the period for charging ($t = 240$ s). We can even pause at every reading and resume when we are ready.

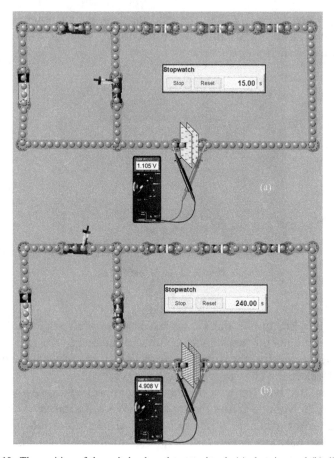

Figure 12.10. The position of the switch when the capacitor is (a) charging and (b) discharging.

12.2.3 Part II: analyzing the results

1. Use Excel to graph the voltage across the capacitor (ΔV_c) vs time (t). We must get a nonlinear graph. The graph displays two parts: an increasing part of the voltage approaching a maximum voltage followed by a decreasing voltage approaching zero voltage. The increasing part is when the capacitor is charging, and the decreasing part is when it is discharging. We recall that when the capacitor is charging, the voltage is given by

$$\Delta V_c(t) = \Delta V_{\max}(1 - e^{-t/\tau}), \tag{12.14}$$

and when it is discharging it is given by

$$\Delta V_c'(t') = \Delta V_o' e^{-t'/\tau'}, \tag{12.15}$$

$\Delta V_o'$ is the voltage across the capacitor just before closing the second switch.

2. Next, we will determine the capacitive time constant using the half-life time ($t_{1/2}$) for the charging period and the discharging period. The $t_{1/2}$ is when it

Table 12.1. The measured potential difference across the capacitor.

Time, t (in seconds)	Voltage across the capacitor, ΔV_c (in volts)
0	0
240	

(*Continued*)

Table 12.1. (*Continued*)

Time, t (in seconds)	Voltage across the capacitor, ΔV_c (in volts)
480	

takes for the voltage across the capacitor to reach half of the measured maximum voltage (when it is charging or discharging). We determine $t_{1/2}$ for charging and discharging by following the procedure listed below:

(a) Read the maximum voltage across the capacitor from your graph.

$\Delta V_{\max} =$

(b) Calculate half of the maximum value.

$\Delta V_{\max}/2 =$

(c) Locate the value for $\Delta V_{\max}/2$ on your graph. We must be able to locate two points on our graph. The first is in the charging part, and the second is in the discharging part of the graph. Read the time from the time axis for these two points.

Charging, $t_1 =$

Discharging, $t_2 =$

(d) Find the half-life time for the charging and discharging.

Charging, $t_{1/2} = t_1 =$

Discharging, $t'_{1/2} = t_2 - 240 \sec =$

3. Using the half-life times, we determined for charging and discharging, calculate the capacitive time constant.

Charging, $\tau = \dfrac{t_{1/2}}{\ln 2} =$

Discharging, $\tau' = \dfrac{t'_{1/2}}{\ln 2} =$

4. Calculate the theoretical value for the capacitive time constant

$\tau = R_{\mathrm{eq}}C =$

5. Sketch of the graph on the graph paper provided in figure 12.11.

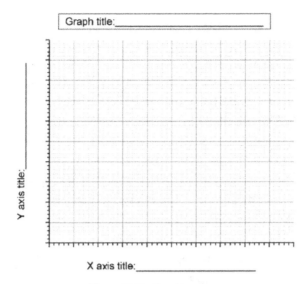

Figure 12.11. Graphing page.

6. Optional: by linearizing the discharging voltage vs time data, make a linear graph using Excel. Find the equation for the best-fit to the linear graph and determine the capacitive time constant.

Discharging, $\tau' =$

12.2.4 Result and conclusion

1. Did all the experimentally determined capacitive time constants agree with the theoretical value within the experimental uncertainties? If not, explain what causes the disagreement.

2. Write a brief overview of what we have accomplished and concluded in this activity.

12.3 Real lab: *capacitors and RC circuits*

12.3.1 Objectives

The objectives of this lab are
- To build an RC circuit with a switch and DC power supply.
- To learn the voltage's behavior across the capacitor when charging and discharging in the dc RC circuit.

12.3.2 Supplies

The supplies we need for this activity are alligators, cables, resistor ($R = 470K\Omega$), capacitor ($C = 100\mu F$), breadboard, and a switch. You also need a multimeter to measure voltage and resistance.

12.3.3 Part I: assembling a circuit with a switch

Measure the resistances of the resistor and the capacitor and record the values. Explain why the resistance for the capacitor reads infinity.

Using the resistor and the capacitor we will build the RC circuit shown in figure 12.12.

1. Using the multimeter set the power supply voltage to $\Delta V_s = 5$ V. After you set the voltage turn the power off and disconnect the multimeter from the power supply.

    You can try to build the circuit in the diagram without following the instructions below (if you like a challenge...).

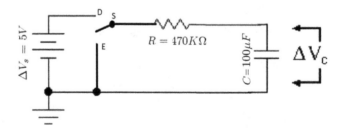

Figure 12.12. The RC circuit with a switch and DC power supply.

2. Connect alligators to the switch, as shown in figure 12.13

Figure 12.13. The switch connected to three alligators.

3. Connect the resistor and the capacitor in series on the breadboard (see figure 12.14 as an example).

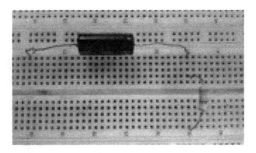

Figure 12.14. Connecting the resistor and the capacitor on the breadboard.

4. Using a red cable, connect the red lead of the power supply to the 'D' lead of the switch (see figure 12.13).
5. Use two cables to connect the S lead of the switch to the resistor and the switch's E lead to the end of the capacitor not connected to the resistor.
6. Using another cable, connect the power supply's black lead to the end of the capacitor not connected to the resistor (Note: there are two cables connected to this end of the capacitor).

   Very Important!!! Make sure the arrows on the capacitor are pointing along the flow direction of the source current (see figure 12.15).

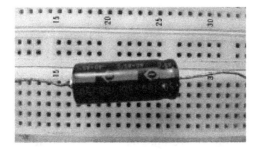

Figure 12.15. The arrows show the direction of the current flow.

12.3.4 Part II: the voltage across the capacitor

1. Connect the multimeter across the capacitor to get ready for measuring the voltage across it. We will measure the voltage across the capacitor when it is charging and discharging. (Remember, for the multimeter, connect the red lead to the positive side and the black lead to the opposing sides of the capacitor.)
2. Set the clock to zero to get ready for reading the voltage from the multimeter. We will read every 15 s for 8 min. The first 4 min is when the capacitor is charging, and the next 4 min is when it is discharging.

   When the switch is at D, and the power supply is on, the capacitor is charging. When it is at E, the capacitor is disconnected from the power supply, and it is discharging (see figure 12.16).

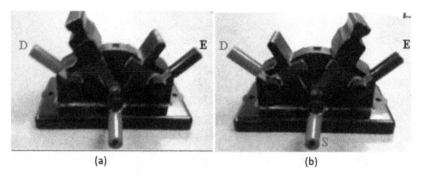

Figure 12.16. The position of the switch when the capacitor is (a) charging and (b) discharging.

3. Now we are about to get started recording the voltage reading on the multimeter. For the first 4 min, the switch stays at D. At exactly $t = 4$ min, the switch is thrown to E and stays at E for the next 4 min. During the entire 8 min, we read the voltage every 15 s.
4. Throw the switch to D, turn the power on, and start the clock running simultaneously. Record the voltage every 15 s in table 12.2. At $t = 4$ min (240 s), throw the switch to E and keep recording the voltage every 15 s until $t = 8$ min (480 s).
5. Estimate the uncertainties for the measurements and calculate the fractional uncertainties.

12.3.5 Part III: analyzing the results

1. Use Excel to graph the voltage across the capacitor (ΔV_c) vs time (t). We must get a nonlinear graph. The graph displays two parts: an increasing part

Table 12.2. The measured potential difference across the capacitor.

Time, t (in seconds)	Voltage across the capacitor, ΔV_c, (in volts)
0	0
15	
30	
45	
60	
75	
90	
105	
120	
135	
150	
165	
180	
195	
210	
225	
240	
255	
270	
285	
300	
315	
330	
345	

(*Continued*)

Table 12.2. (*Continued*)

Time, t (in seconds)	Voltage across the capacitor, ΔV_c, (in volts)
360	
375	
390	
405	
420	
435	
450	
465	
480	

of the voltage approaching a maximum voltage and a decreasing voltage approaching the zero voltage. The increasing part is when the capacitor is charging, and the decreasing part is when it is discharging. We recall that when the capacitor is charging, the voltage is given by

$$\Delta V_c(t) = \Delta V_{max}(1 - e^{-t/\tau}), \qquad (12.16)$$

and when it is discharging it is given by

$$\Delta V_c'(t') = \Delta V_o' e^{-t'/\tau'}, \qquad (12.17)$$

where $\Delta V_o'$ is the voltage across the capacitor just before the switch is thrown to E (i.e. at t = 4 min).

2. Next, we will determine the capacitive time constant using the half-life time ($t_{1/2}$) for the charging period and the discharging period. The $t_{1/2}$ is the time it takes for the voltage across the capacitor to reach half of the measured maximum voltage (when it is charging or discharging). We determine $t_{1/2}$ for charging and discharging by following the procedure listed below:

   (a) Read the maximum voltage across the capacitor from your graph.

   (b) Calculate half of the maximum value $\Delta V_{max}/2$.

(c) Locate the value for $\Delta V_{max}/2$ on your graph. You must be able to locate two points on your graph. The first is in the charging part and the second is in the discharging part of the graph. Read the time from the time axis for these two points.

(d) Find the half-life time for charging and discharging.

3. Using the half-life times you determined for charging and discharging, calculate the capacitive time constants along with the uncertainties.

4. Calculate the theoretical value for the capacitive time constant.

5. Sketch the graph on the graph paper provided in figure 12.17.

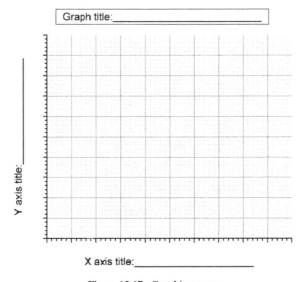

Figure 12.17. Graphing page.

6. Optional: by linearizing the discharging voltage vs time data, make a linear graph using Excel. Find the equation for the best-fit to the linear graph and determine the capacitive time constant along with the uncertainties.

12.3.6 Result and conclusion

1. Did all the experimentally determined capacitive time constants agree with the theoretical value within the experimental uncertainties? If not, explain what causes the disagreement.

2. Write a brief overview of what we have accomplished and concluded in this activity.

IOP Publishing

Virtual and Real Labs for Introductory Physics II
Optics, modern physics, and electromagnetism
Daniel Erenso

Chapter 13

Oscilloscope and fast time constant

In the last chapter, we studied a dc RC circuit when charging and discharging by integrating a switch. Since the capacitive time constant for such an RC circuit is slow, we measured the potential difference across the capacitor as a function of time using a multimeter and a stopwatch. However, an RC circuit could have a fast capacitive time constant, and it is impossible to get an accurate reading for the potential difference using a multimeter. In such a case, we usually use an oscilloscope to display the potential difference vs time. This chapter introduces us to an oscilloscope and function generator and a dc RC circuit with a fast capacitive time constant. After a brief introduction to a function generator and an oscilloscope, we perform virtual and real labs. In the virtual lab, we use the PhTH simulation to generate different functions such as square, TTL (transistor–transistor logic), and sinusoidal functions in an electrical circuit. The real lab has two parts. The first part introduces the basic features of an oscilloscope and a function generator. In the second part, we measure a fast capacitive time constant in an RC circuit using an oscilloscope and a TTL signal from a function generator.

13.1 Basic theory

Function generator
A function generator is usually a piece of electronic test equipment or software used to generate different electrical waveforms over a wide range of frequencies. Some of the most common waveforms produced by the function generator are the sine wave, square wave, triangular wave, and sawtooth shapes. These waveforms can be either repetitive or single-shot (which requires an internal or external trigger source). Integrated circuits (IC) used to generate waveforms may also be described as function generator ICs.

Another feature included on many function generators is the ability to add a DC offset.

Oscilloscope

An oscilloscope is a type of electronic test instrument that graphically displays varying signal voltages, usually as a two-dimensional plot of one or more signals as a function of time. Other signals (such as sound or vibration) can be converted to voltages and displayed. Oscilloscopes display an electrical signal change over time, with voltage and time as the Y- and X-axes, respectively, on a calibrated scale. The waveform can then be analyzed for properties such as amplitude, frequency, and others. Modern digital instruments may calculate and display these properties directly. Originally, the calculation of these values required manually measuring the waveform against the scales built into the instrument's screen. The oscilloscope can be adjusted to observe repetitive signals as a continuous shape on the screen. A storage oscilloscope can capture a single event and display it continuously, so the user can observe events that would otherwise appear too briefly to see directly.

13.2 Virtual lab I: *function generator and voltage chart*

13.2.1 Introduction

The objectives of this virtual lab are

- To study the difference between a voltage chart (oscilloscope) and a voltmeter.
- To learn how to generate square and TTL voltage signals and study the difference between these two types of signals.
- To get introduced to an ac voltage source that generates sinusoidal voltage signals.
- To learn how to read amplitude and period.
- To study the relation between period and frequency.

To this end, we will be using the same simulation kit we used in the previous lab. To open this simulation kit go to/click https://phet.colorado.edu/en/simulation/legacy/circuit-construction-kit-ac. You should see what you saw in the previous virtual lab shown in figure 13.1. Download and open the file and it leads us to the simulation window shown in figure 13.2.

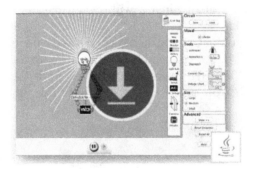

Figure 13.1. Circuit construction kit (AC + DC), Virtual Lab.

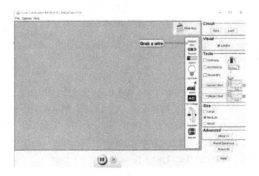

Figure 13.2. Opened circuit construction kit.

13.2.2 Part I: voltmeter vs voltage chart (oscilloscope)

1. Using the dc voltage (battery), a switch, a resistor, and wires, construct the circuit shown in figure 13.3. Keep the switch off until we are instructed to turn it on.

Figure 13.3. A battery, a switch, and a resistor.

2. Set the resistance $R = 5\ \Omega$ and the source voltage, $\Delta V_s = 5V$. For this kit, to change the source voltage or the resistance, we right-click on the battery or the resistor, and a small window will open where we can make the changes we want to make.

3. Click on the 'voltmeter' and hook up the red and black leads across the resistor. Click on the 'voltage chart' and hook up the red and black leads across the resistor. We must read zero on both the 'voltmeter' and the 'voltage chart' if you have kept the switch off (see figure 13.4).

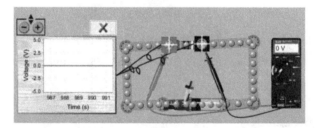

Figure 13.4. Measuring the voltage across the resistor (which is equal to the source voltage) using a voltmeter and 'voltage chart' (oscilloscope).

4. *Setting up 'the voltage charts'*: as we can see in figure 13.5, the vertical axis has four divisions, and the horizontal axis has two divisions. To set this up for the vertical axis, we can click on '−' or '+'. On the other hand, to set the time axis, click on play and stop just before the dotted vertical line reaches the voltage axis. Then use the slow play to line up the dotted vertical line with the voltage axis (y-axis). *If we have carried all the steps above successfully, then we can proceed to the next step.*

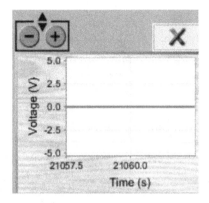

Figure 13.5. The setting for the voltage chart.

5. What are the voltage per division (V/div) for the vertical axis and time per division (t/div) for the horizontal axis?

6. Now turn the source switch on. We must see the voltage across the resistor $\Delta V_R(t)$, which must be equal to the source voltage $\Delta V_s(t)$ being displayed as a function of time t. Switch the polarities for both the voltmeter and the voltage chart. Briefly discuss our observation about the source voltage, the real current, and the conventional current.

7. Stop the simulation, turn the switch off, remove the voltmeter, and switch back the 'voltage chart' polarities. Click on 'stopwatch' to clock the time for the next part.

13.2.3 Part II: A square, TTL, and sinusoidal voltage signals

In this part of the lab, we are interested in generating three types of signals (square, TTL, and sinusoidal signals) and study the similarities and differences in these signals' properties.

1. It is always important to set-up the 'voltage chart' for easier reading of voltage and time. So repeat the procedure described in step 4 in Part I. Reset the stopwatch (see figure 13.6).

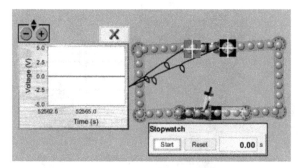

Figure 13.6. The voltage across the resistor with the switch off.

2. **Function one**
   (a) Turn the switch on and start the stopwatch.
   (b) Run the simulation using the slow play until the stopwatch reads 1.02 s. Pay attention to how the charges move in the circuit.
   (c) Reverse the source current (the battery). (To reverse the source current right-click on the battery and select reverse.) Resume the simulation till the stopwatch reads 2.04. Here also pay attention to the charges.
   (d) Reverse the source current and resume the simulation for the stopwatch readings: 3.06 s, 4.08 s and 5.10 s. At the end of 5.10 s, we should expect to see, for example, figure 13.7.
3. Identify the signal (function) type you have just generated and describe its properties in terms of the current in the circuit.

4. Read the amplitude and the period in divisions from the 'voltage chart' and find the corresponding values in volts and seconds, respectively. Using the period, determine the frequency. Record the values at the appropriate place in table 13.1.

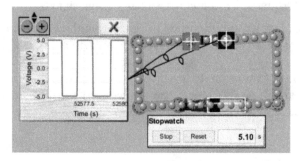

Figure 13.7. The voltage across the resistor with the switch on and the polarity flipped.

13-6

Table 13.1. Measured amplitude and period.

Signal	Amplitude		Period		Frequency in Hz
	In DIV	In volts	In DIV	In seconds	
Square					
TTL					
Sine					

5. **Function two**
   (a) Do the procedure described in step 1 of Part II above.
   (b) Turn the switch on and start the stopwatch.
   (c) Run the simulation using the slow play for 1.02 s. Pay attention to how the charges move in the circuit.
   (d) Turn the switch off and repeat step (c). (Here also pay attention to the charges.)
   (e) Repeat steps (c) and (d) until the stopwatch reads 4.08 s. We should expect to see, for example, figure 13.8.

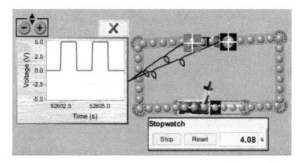

Figure 13.8. Voltage across the resistor with the switch on and off.

6. Identify the signal (function) type you have just generated and describe its properties in terms of the current in the circuit.

7. Repeat step 4.
8. Briefly explain the difference between a square wave and a TTL signal.

Make sure the switch is off before we go to the next step.

9. **Function three**
   (a) Remove the battery, replace it with ac voltage, and set the amplitude to 5 V. Also, set-up the voltage chart (step 1 of Part II).
   (b) Turn the switch on and reset the stopwatch.
   (c) Simulate for about 10.0 s. At the end of this step, we should expect to see something that resembles figure 13.9.
10. Identify the signal (function) type that the ac voltage generates and describe

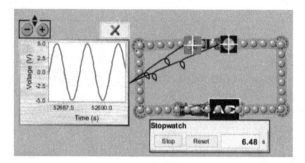

Figure 13.9. Voltage across the resistor with an ac source with the switch on.

its properties in terms of the current in the circuit.

11. Repeat step 4. Right-click on the ac voltage and read the frequency.

12. Find the percent difference between the frequency we determined from the period (experimental) and what you read from the ac voltage (theoretical).

13. Based on the percent difference, can we say the experimental result agrees with the theoretical within the uncertainties'? What could be the primary source of our uncertainties?

13.2.4 Result and conclusion

Write a brief overview of what was accomplished and concluded in this activity.

13.3 Virtual lab II: *fast capacitive time constant*

13.3.1 Introduction

The objectives of this virtual lab are
- To construct and study the properties of an RC circuit with a fast capacitive time constant.
- To experimentally determine the capacitive time constant in a dc RC circuit when a capacitor is charging and discharging faster.

We will be using the same simulation kit we used in the previous lab. To open this simulation kit go to/click https://phet.colorado.edu/en/simulation/legacy/circuit-con-struction-kit-ac. You should see what you saw in the previous virtual lab shown in figure 13.10. Download and open the file, and it leads us to the simulation window shown in figure 13.11.

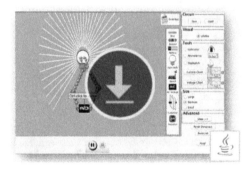

Figure 13.10. Circuit construction kit (AC + DC), Virtual Lab.

Figure 13.11. Opened circuit construction kit.

13.3.2 Part I: Charging the capacitor

1. Using the dc voltage (battery), two switches, two resistors, a capacitor, and wires construct the circuit shown in figure 13.12.

Figure 13.12. Dc RC circuit.

2. Set the resistance $R_1 = R_2 = R = 5\,\Omega$, the capacitance $C = 0.05F$, and the source voltage, $\Delta V_s = 5$ V.
3. Click on the 'voltage chart' and hook up the red and black leads across the capacitor. Once again, click on the 'voltage chart' and hook up the red and black leads across the resistor on the right side of the circuit. Click on 'stopwatch' (see figure 13.13).

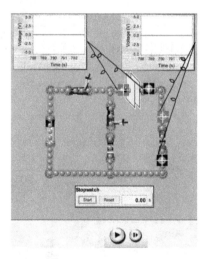

Figure 13.13. Measuring the voltage across the capacitor and resistor in a dc RC circuit.

4. Keep both switches off, and if we accidentally turned it on and the current is flowing in the circuit, turned it back off and discharge the capacitor.
5. *Setting up 'the voltage charts'*: as we can see in figure 13.14, the vertical axis has four divisions, and the horizontal axis has two divisions. To set this up for the vertical axis, we can click on '-' or '+'. On the other hand, to set the

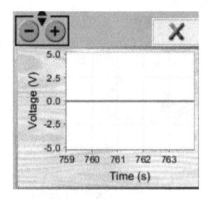

Figure 13.14. The setting for the voltage chart.

time axis, click on play and stop just before the dotted vertical line reaches the voltage axis. Then use the slow play to line up the dotted vertical line with the voltage axis (y-axis).

If we have carried all the steps above successfully, then we can proceed to the next step.

6. What are the voltage per division ($V/$ div) for the vertical axis and time per division ($t/$ div) for the horizontal axis?

7. Now turn the source switch on (the one on the horizontal wire in figure 13.13) and click play. You must see the voltage across the capacitor $\Delta V_C(t)$ and the voltage across the resistor $\Delta V_R(t)$ being displayed as a function of time t. When we stop the simulation, for example, after about 1 s, we see the traces for $\Delta V_C(t)$ and $\Delta V_R(t)$ that resembles the graphs in figure 13.15.

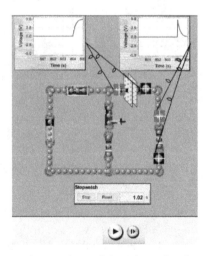

Figure 13.15. The voltage across the capacitor and the resistor after the switch is turned on ($t \simeq 1.0s$).

8. Use the fast and slow play buttons till we see full traces for $\Delta V_C(t)$ and $\Delta V_R(t)$ (for example, see in figure 13.16).

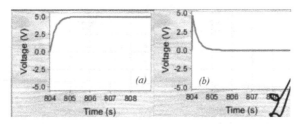

Figure 13.16. The voltage across (a) the capacitor and (b) the resistor as function of time during charging.

9. Figure 13.16(a) and (b) show that, at the initial time, $\Delta V_C(t)$ is equal to zero ($\Delta V_C = 0$) and $\Delta V_R(t)$ is equal to the source voltage ($\Delta V_R = \Delta V_s = 5$ V). But as time goes on $\Delta V_R(t)$ approaches to zero, and $\Delta V_C(t)$ reaches close to the source voltage. Give a brief explanation of why this is happening. What does this generally tell us about capacitors' property in a dc RC circuit at $t = 0$ and $t = \infty$?

10. At any given point of time after the switch is turned on, what relation can we establish between the source voltage ΔV_s and the total voltage across the capacitor and the resistor ($\Delta V_C(t) + \Delta V_R(t)$)?

11. Using the full display for $\Delta V_C(t)$ that you see on the screen, find the values for the quantities listed in table 13.2.

Table 13.2. Voltage across the capacitor.

Voltages	In DIV	
Maximum voltage across the capacitor		
Half of the maximum voltage		
Times	**In DIV**	**In Seconds**
$t_{1/2}$		

12. Find the theoretical τ_{theo} and experimental τ_{exp} values for the capacitive time constant in seconds and record the values in table 13.3.

Table 13.3. Capacitive time constant.

$\tau_{\text{theo}} = R_e C$	
$\tau_{\text{exp}} = \frac{t_{1/2}}{\ln 2}$	

13. What is the equivalent resistance, R_e, of the circuit?

14. Using the experimental value for the capacitive time constant τ and the equivalent resistance, R_e, find the capacitance for the capacitor.

15. Find the percent difference between the theoretical and experimental values for the capacitance. Does the result for the experimental values agree with the theoretical, within the experimental uncertainties? If not, explain why?

13.3.3 Part II: Discharging the capacitor

Note: it is essential to keep everything the same in order to do the next part. (We should have on our screen what we see in figure 13.17.)

1. *Do not play the simulation until we are instructed to do so.* Turn the first switch off (what does this do to the source voltage?) and then turn on the second switch. Find the equivalent resistance of the circuit, R_e?

2. Reset the stopwatch. Move the black lead of the voltage chart hooked across the first resistor to the top end of the second resistor (the resistor in the middle of the circuit) to measure the total voltage across the two resistors (see figure 13.18).

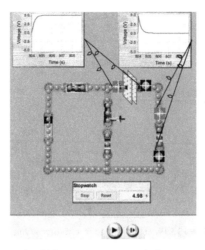

Figure 13.17. What you should have on your screen before you move to the next part.

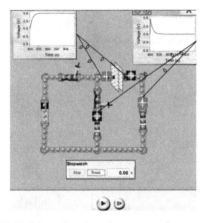

Figure 13.18. Measuring the voltage across the two resistors.

3. Run the simulation (play) and stop it after about 1.0 s. You must see what you see in figure 13.19.
4. Use fast and slow play to get a full display for $\Delta V_C'(t)$ and $\Delta V_R'(t)$ (see figures 13.20(a) & (b)).
5. What relation can we establish between $\Delta V_C'(t)$ and $\Delta V_R'(t)$, where $\Delta V_R'(t)$ is the total voltage across the resistors?

6. Using the full display for $\Delta V_C'(t)$, find the values for the quantities listed in table 13.4.
7. Find the theoretical and experimental values for the capacitive time constant in seconds. Record the values in table 13.5.

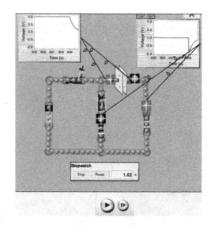

Figure 13.19. The simulation run after about 3 s.

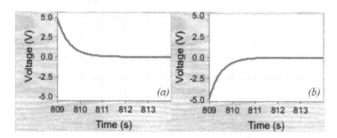

Figure 13.20. The voltage as a function of time across (a) the capacitor and (b) the resistor during charging.

Table 13.4. Measured voltage across the capacitor.

Voltages	In DIV	
Maximum voltage across the capacitor		
Half of the maximum voltage		
Times	**In DIV**	**In Seconds**
$t'_{1/2}$		

Table 13.5. Capacitive time constant.

Theoretical inductive time constant, $\tau_{\text{theo}} = CR_e$	
Experimental inductive time constant, $\tau_{\text{exp}} = \dfrac{t'_{1/2}}{\ln 2}$	

8. Using the experimental value for the capacitive time constant, find the circuit's resistance, R_e.

9. Does the result for the experimental values agree with the theoretical, within the uncertainties. If not, explain why?

13.3.4 Result and conclusion

Write a brief overview of what was accomplished and concluded in this activity.

13.4 Real lab I: *introduction to oscilloscope*

13.4.1 Objectives

The objectives of this lab are
- To gain an understanding of the controls of the oscilloscope.
- To compare a voltmeter with an oscilloscope.
- To gain an appreciation of the oscilloscope as an experimental tool.
- To gain an understanding of the use of a function generator

13.4.2 Supplies

In this activity, like the previous activities, we need alligators, cables, the multimeter, and the power supply (figure 13.21). The new supplies in this activity are the oscilloscope, the function generator, and special cables known as BNC (oscilloscope cables) (see figure 13.22).

Figure 13.21. From left to right: alligators, cables, multimeters, and power supply.

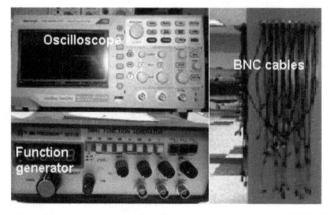

Figure 13.22. The new supplies (oscilloscope, function generator, and BNC cables).

13.4.3 The very basic elements of the oscilloscope

We will learn the basic elements of the oscilloscope shown in figure 13.22.

The oscilloscope screen shows a vertical/horizontal grid of lines on which a voltage vs time graph will be projected for the input voltage signal. Each block on the grid (square of about 1 cm in size) is referred to as a division, abbreviated DIV. The vertical axis on the grid will represent voltage, and the horizontal axis will represent time. We can easily change the vertical position of the zero volts level. We usually use the very bottom grid line or the center grid line as the 0V (or ground) level (see figure 13.23 as an example).

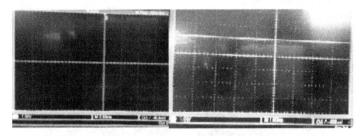

Figure 13.23. The yellow horizontal line shows the 0V ground position. Left: ground set at the very bottom grid line. Right: ground set at 1 division above the center grid line.

- **The vertical positioning knob** (yellow knob marked A in figure 13.24) moves the trace up or down. It shifts the whole graph up or down on the face of the oscilloscope screen.
- **Channel 1 button** (button B) turns on and off the channel 1 signal displayed on the oscilloscope screen.
- **The vertical sensitivity knob** (knob C in figure 13.24) will control the voltage scale on the vertical axis of the grid. Each grid mark (DIV) may represent 1 V

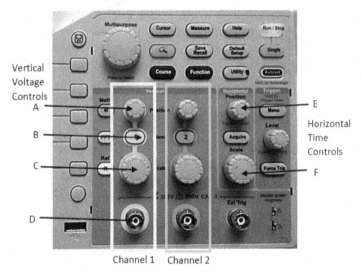

Figure 13.24. The essential control knobs of the oscilloscope

or it may represent 5 V (or some other value). The voltage scale is displayed at the bottom of the oscilloscope screen.

- **The horizontal position knob** (knob E in figure 13.24) moves the grid plot to the right or left. This control is used to position the plot horizontally for our convenience without affecting the vertical 0 V setting.
- **The horizontal sweep-time control** (knob F in figure 13.24) will control the time scale on the horizontal axis. This control will tell us how many seconds correspond to each horizontal division. The time scale is displayed at the bottom of the oscilloscope screen.
- **The channel 1 input connector** (connector marked D in figure 13.24)
- *Input*: an electrical signal can be displayed on the oscilloscope screen by connecting the electrical signal to the oscilloscope **Input** using **BNC** (Oscilloscope cables: figure 13.25).

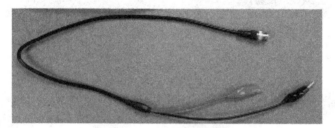

Figure 13.25. BNC cable

13.4.4 Our first oscilloscope trace!

1. Turn on the oscilloscope power by pressing the button on the top right side of the oscilloscope.
2. Position the 0V ground level at the center level on the oscilloscope grid. To set the ground level, press the yellow channel one button (button B), and you will see a yellow horizontal line and voltage and time settings at the bottom of the oscilloscope screen. If we do not see it, we can press the button again. Adjust the vertical (A) and horizontal (E) positioning knobs until the horizontal yellow trace is lined up with and centered on the centerline (the horizontal line with small tick marks).
3. Connect the voltmeter cables to the output terminal of the dc power supply, red-to-red and black-to-black. Into these cables connect channel one input BNC cables of the oscilloscope. (The BNC input cable for the oscilloscope has to be pushed in and then turned fully clockwise until we can feel it click into place.) Turn on the power supply and the voltmeter and turn up the output voltage to about 3.5V. Record the voltmeter reading.
4. Set the vertical sensitivity scale to 1V. How many divisions (DIV) is the horizontal line above the 0V level at the screen center? How many volts does this correspond to? Does this value agree with the voltmeter reading within the experimental uncertainties? (It should!)

5. Now change the vertical sensitivity to 2V. How many divisions is the line above the 0V level now? To how many volts does this correspond?
6. Now change the output of the dc power source to 5V and repeat steps 4 and 5. Record the measurements in table 13.6.

Table 13.6. Voltage measurement by the oscilloscope and the voltmeter.

Power output (voltmeter reading)	Oscilloscope reading at 1V/DIV		Oscilloscope reading at 5V/DIV	
	In DIV	In volts	In DIV	In volts
2 volts				
5 volts				
5 volts (switched polarity)				

13.4.5 A fancier trace and time measurements

1. Turn off the power supply and disconnect the voltmeter and the oscilloscope from the terminals.
2. Locate the function generator (see the function generator's picture at the beginning of the lab) and press the sine-wave output button.
3. Connect the BNC cable to the output connector of the function generator, figure 13.26.
4. Set the AMPL(itude) and the FREQUENCY (both COARSE and FINE) knobs to about the center of their range.
5. Connect the output cables of the function generator to the channel one input cables of the oscilloscope. Check the ground of the oscilloscope to ensure that it is still at the center of the grid as before.
6. Turn on the function generator and press the button for the 1 *Kilohertz* frequency range (this gives frequencies up to about 1 kHz). The digital readout on the function generator will show us the frequency of the signal it

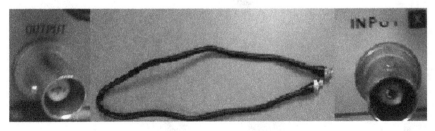

Figure 13.26. Output, BNC cable, and oscilloscope input.

generates. Turn the coarse and fine frequency control knobs until the frequency is close to 1300 Hz.

7. We should see a trace on the oscilloscope—the trace will most likely be jumbled. First, set the vertical sensitivity control knob (knob C) so that the displayed signal's amplitude is as large as possible but still fits entirely on the oscilloscope grid.

8. We can now set the horizontal axis scale. Change the sweep-time setting knob (knob F) so that you can see at least one full period of the sine wave, but not more than two or three periods. We want to measure the amplitude and the period of this wave, so getting one complete period of the wave as large as possible on the grid will make our measurements easier and precise. The waveform's amplitude is the voltage from the central (0 V) axis to the very top (or bottom) of the trace. The period is the time from one point on the wave to the next equivalent point (see figure 13.27 as an example).

9. Make and label a sketch of the oscilloscope trace.

10. Determine the number of divisions for the wave's amplitude and period (what uncertainty would you estimate with these determinations?). Then convert these numbers to voltage and seconds by using the vertical sensitivity and sweep-time settings read from the bottom of the oscilloscope grid.

11. Use the period measurement from the oscilloscope to determine the voltage signal's frequency from the function generator. Does the result agree with the function generator display within the experimental uncertainties? (It should!)

12. Press the square-wave button on the function generator; the sine wave will be replaced by a square wave (see figure 13.28). Repeat steps 9-11.

13. Examine the TTL output of the function generator in the same manner as the previous step. We will have to move the BNC cable from the OUTPUT to the TTL/CMOS terminal on the function generator. We will see a signal that looks like figure 13.29. How does the TTL signal differ from the square-wave signal studied in the previous step? Explain.

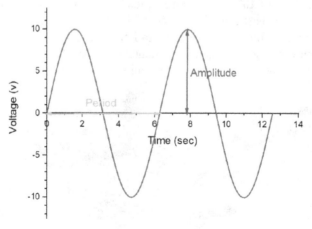

Figure 13.27. Period and amplitude of a wave.

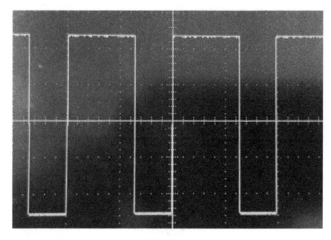

Figure 13.28. Square wave

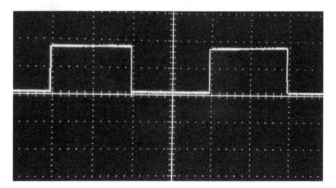

Figure 13.29. A TTL signal.

Table 13.7. Quantities describing the sine and square waves, and TTL signal.

Signal	Amplitude		Period		Frequency in Hz
	In DIV	In volts	In DIV	In seconds	
Sine					
Square					
TTL					

14. Record the values for the quantities listed in table 13.7.
15. Turn everything off and remove the BNC cable from the function generator and the oscilloscope. Please keep all the supplies we used as we will be using them for the next activity!

16. Recall how the switch was used in the previous activity (RC circuit). With the switch 'ON', we included a dc voltage in the circuit. With the switch 'OFF', the input was grounded. The TTL signal in this activity has two characteristic voltages—we can call these voltages the 'ON' (the higher) and 'OFF' (the lower) voltages. What is the value of the OFF-voltage? (In the next activity, we will see how the TTL signal output of the function generator can be used as a swift switch in a circuit.)

13.5 Real lab II: *fast capacitive time constant RC circuit*

13.5.1 Objectives

The objectives of this lab are
- To use an oscilloscope to determine the fast capacitive time constant.
- To use the TTL signal from a function generator as a switch.

13.5.2 Supplies

The additional supply we need is a capacitor with capacitance $C = 0.01\mu F$ and a resistor with resistance $R = 15K\Omega$. Measure the resistance of the resistor and the capacitor and record the values.

13.5.3 Fast capacitive time constant RC circuit

Using the resistor and the capacitor, we will build the RC circuit shown in figure 13.30.

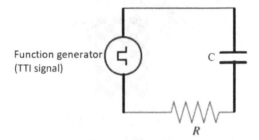

Figure 13.30. RC circuit.

1. Connect the resistor and the capacitor in series on the breadboard.
2. Connect a BNC cable of the type shown in figure 13.31 to the TTL/COMS output jack's function generator.
3. Connect the BNC cable's red lead to the resistor end that is not connected to the capacitor (point D).
4. Connect the BNC cable's black lead to the capacitor end that is not connected to the resistor (point A).

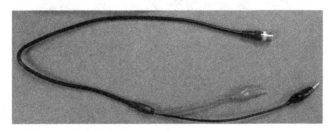

Figure 13.31. Recommended BNC cable.

5. Turn on the function generator. Push the button for 1 kHz. Get a frequency reading of about 1 kHz by adjusting the COARSE and FINE knobs.
6. Set the TTL signal's amplitude to the maximum value by turning the AMPL(itude) knob of the function generator.
7. Turn on the oscilloscope and set the ground at the time axis as shown in figure 13.32 for the channel that we will be using.

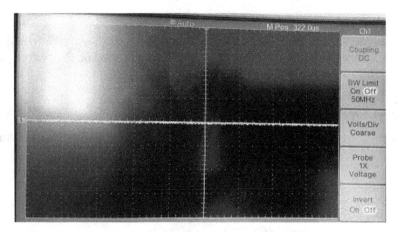

Figure 13.32. Ground set at the middle of the screen.

8. Connect another BNC cable of the type shown in figure 13.31 to the oscilloscope, whichever channel you decided to use (CH1 or CH2).
9. Connect the BNC cable's red lead to the resistor end that is not connected to the capacitor (point D).
10. Connect the BNC cable's black lead to the capacitor end that is not connected to the resistor (point A).
11. If the circuit is built right, we must be able to see the TTL signal in figure 13.33.

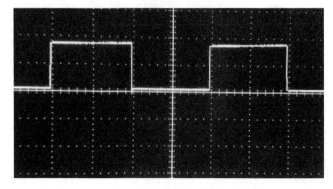

Figure 13.33. TTL Signal

12. If you do not see this signal you may want to check your circuit. Otherwise proceed to the next step.
13. From the signal that we see on the screen, find the list of quantities in table 13.8.
14. Connect the oscilloscope leads so that it reads the voltage across the capacitor ΔV_c. We must see the voltage of the capacitor that is being charged and discharged as the TTL signal is on and off (figure 13.34).
15. Superimposed on the voltage across the capacitor (in figure 13.34), sketch the TTL voltage and identify the charging and discharging voltage.
16. Move the ground to the bottom of the screen (see figure 13.35)
17. It would be best to have at least something that looks like figure 13.36. (Do not move this trace vertically! That would change the ground!)
18. From this trace of the discharging voltage that we see on the screen, find the values listed in table 13.9.

Table 13.8. Quantities describing the TTL signal.

Voltages	In DIV	In volts
Maximum voltage of the TTL signal ('On' voltage)		
Minimum voltage of the TTL signal ('Off' voltage)		
Times	**In DIV**	**In seconds**
Time over which the TTL signal is 'On'		
Time over which the TTL signal is 'Off'		

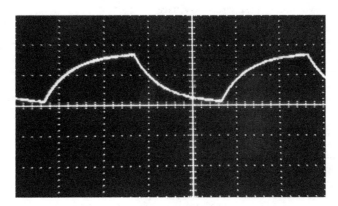

Figure 13.34. The voltage across the capacitor.

13-27

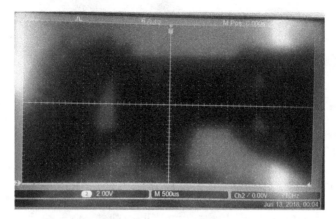

Figure 13.35. The new ground.

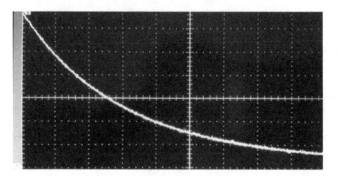

Figure 13.36. The trace of the discharging capacitor voltage.

Table 13.9. Quantities describing the discharging voltage.

Voltages	In DIV	In volts
Maximum voltage across the capacitor		
Half of the maximum voltage		
Times	**In DIV**	**In seconds**
Time at which the voltage reduces to half of the maximum ($t_{1/2}$)		

19. Estimate the uncertainties for the measurements.
20. Find the theoretical and experimental values for the capacitive time constant in seconds.

13.5.4 Result and conclusion

Write a brief overview of what was accomplished and concluded in the activities you have done.

IOP Publishing

Virtual and Real Labs for Introductory Physics II
Optics, modern physics, and electromagnetism
Daniel Erenso

Chapter 14

The magnetic field

A moving charge creates an electric current, and an electric current produces a magnetic field. This chapter introduces us to a magnetic field through virtual and real labs. We begin by introducing the fundamental theories associated with the magnetic field vector of a long current-carrying wire and a solenoid, and the magnetic force on a moving charge and current-carrying wire due to an external magnetic field. In the virtual lab component, we use the PhTH simulation to study the Earth's magnetic field and a bar magnet. We then map the magnetic field vector for a bar magnetic, a long current-carrying wire, and a solenoid in the real lab. We also study the properties of ferromagnetic materials when exposed to an external magnetic field.

14.1 Basic theory

A moving charge generates an electric current, and an electric current produces a magnetic field.

Long current-carrying wire

Consider a long wire that carries a current I flowing up figure 14.1(a) or down (b). The magnetic field (magnitude) due to this current at a distance d from the wire is given by

$$B = \frac{\mu_0 I}{2\pi d},\qquad(14.1)$$

where

$$\mu_0 = 4\pi \times 10^{-7}\frac{\text{T}\cdot\text{m}}{\text{A}},\qquad(14.2)$$

is known as the magnetic permeability of a free space, MKS unit of magnetic field is Tesla. The direction is determined using the *right-hand rule I: extend your right-hand*

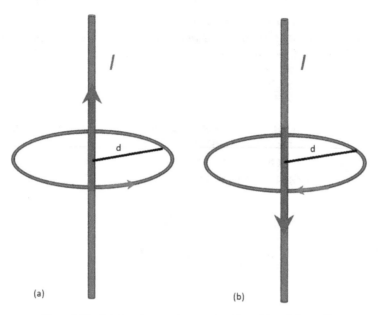

Figure 14.1. A long wire carrying a current I up (a) and down (b).

thumb along the direction of the current and your fingers curl around the direction of the magnetic field.

Vectors cross product

The cross product of two vectors $\vec{A}$ and $\vec{B}$ (see figure 14.2) gives a third vector $\vec{C}$

$$\vec{C} = \vec{A} \times \vec{B}. \tag{14.3}$$

The magnitude of the vector $\vec{C}$ is given by

$$C = AB \sin(\theta), \tag{14.4}$$

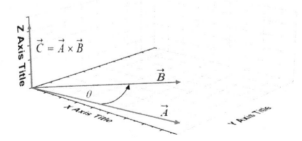

Figure 14.2. The cross (vector) product of two vectors.

14-2

and the direction is given by the *right-hand rule II: point your right-hand thumb along the direction of the first vector A and extend the remaining fingers along the direction of the second vector B, then the palm points along the direction of the vector C.*

Magnetic force

A moving charge and current carrying wire can experience a magnetic force in a region where there is a magnetic field.

(a) *Moving charge*: a charge Q moving with a velocity $\vec{v}$ in a region of magnetic field $\vec{B}$, the magnetic force $\vec{F}$ is given by

$$\vec{F} = Q(\vec{v} \times \vec{B}).\tag{14.5}$$

The magnitude of this force is determined using,

$$F = |Q|\, vB \sin(\theta),\tag{14.6}$$

where θ is the angle between the velocity and the magnetic field vectors. The direction is determined using *right-hand rule II*.

(b) *Current-carrying wire*: a wire of length l carrying a current I in a region of magnetic field $\vec{B}$, the magnetic force $\vec{F}$ is given by

$$\vec{F} = I(\vec{l} \times \vec{B}),\tag{14.7}$$

where the direction of the vector $\vec{l}$ is the direction of the current. The magnitude is

$$F = IlB \sin(\theta),\tag{14.8}$$

where θ is the angle between the current and the magnetic field directions. The direction of force is determined using *right-hand rule II*.

Magnetic dipole

A magnetic dipole is a loop carrying a current I. The north pole (field lines emanating from) and south pole (field lines moving in towards) are determined using *right-hand rule III: rap fingers around the direction of the current flow (see figure 14.3) in the loops, then the thumb points along the direction of the magnetic field (the north pole).*

Ferromagnets: materials that become a magnet when exposed to external magnetic fields.

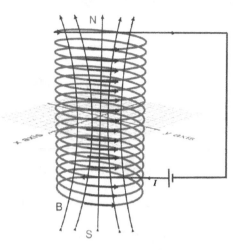

Figure 14.3. The current and magnetic field in a solenoid.

14.2 Virtual lab: *magnetic field*

14.2.1 Introduction

The objectives of this virtual lab are
- To study the properties of a magnetic field.
- To determine the magnetic field lines for a bar magnet.
- To study the difference between geographic and magnetic poles for our planet Earth.

To this end, go to/click on http://phet.colorado.edu/en/simulation/legacy/magnet-and-compass to open the simulation kit shown in figure 14.4. Download, open the file, and it directs us to the magnetic field simulation window shown in figure 14.5.

Figure 14.4. Magnetic field simulation.

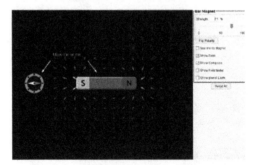

Figure 14.5. Virtual lab for a magnetic field.

14.2.2 Part I: a bar magnet

1. Which tip is the north pole of the compass (the red or white tip)? How do we know?

2. Flip the polarity of the bar magnet by clicking 'flip polarity'. What happened to the compass? Explain!

3. Which tip is the south pole of the compass (the red or the white tip)?

4. Move the compass around the bar magnet about an inch away and observe the compass's white and red tips.

5. Imagine we placed this bar magnet at the center of a clean sheet of paper on a table, and we draw a small arrow on the paper that points from the south to the north pole of the compass as we move the compass around. Show these arrows in figure 14.6 and label the bar magnet's north and south poles.

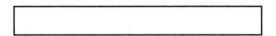

Figure 14.6. The magnetic field directions of a bar magnet.

6. Check the box 'See inside magnet', draw a small arrow on the paper that points from the south to the north pole of the compass in figure 14.12. By tracing a curve that connects these arrows show the magnetic field lines for a bar magnet.

   The field-line pattern for the bar magnet is fundamental. It describes what is known as a dipole-field pattern. Anything acting like a magnetic dipole will have this pattern of field lines.

7. Uncheck the box 'Show compass' and check the box 'Show Field Meter'. Move the field meter around the white curve shown in figure 14.7. Record evenly distributed 10 data points for the x and y components of the magnetic field in table 14.1.

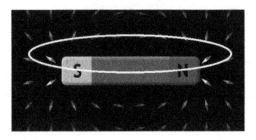

Figure 14.7. A bar magnet.

8. Use Excel to make a scatter plot for the data in table 14.1. Sketch the data points in the space provided below. Draw the best fitting curve to these data

Table 14.1. The x and y components of the magnetic field.

B_x (in G)	B_y (in G)

points. Does this curve predict the magnetic field lines we determined in step 6?

9. In our study of the electric field, we have seen that the electric field of a charge is inversely proportional to the charge's distance from the point of interest. What kind of relationship can we predict for magnetic field and distance for a magnet, at least for the bar magnet we studied. We can do measurements for the magnetic field at different distances from the bar magnet.

14.2.3 Part II: Earth's magnetic field

1. Check the box, 'Show planet earth'. Do we see a giant bar magnet inside the Earth? Yes, the Earth is like one giant bar magnet. It has north and south magnetic poles like a bar magnet. We will find these magnetic poles using the compass that we already know its north and south poles.

2. Move the compass to the geographic north pole (the pole that typically referred to as north); which tip of the compass is attracted to the geographic north pole?

3. Move the compass to the geographic south pole (the pole that commonly referred to as south); which tip of the compass is attracted to the geographic south pole?

4. Based on the observation, briefly discuss Earth's magnetic field.

5. Do steps 4 & 5 in Part I and trace Earth's magnetic field lines in figure 14.8. Label the geographic and magnetic poles in the sketch.

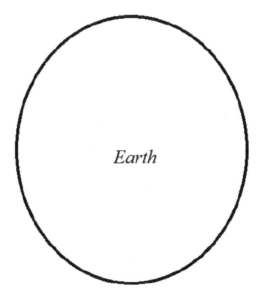

Figure 14.8. Earth's magnetic field lines.

14.2.4 Result and conclusion

Write a brief overview of what we accomplished and concluded in this activity.

14.3 Real lab: *magnetic field*

14.3.1 Objectives

The objectives of this activity are
- To study the magnetic field of a bar magnet.
- To map the magnetic field of a long current-carrying wire.
- To study the magnetic field of an electromagnet (solenoid).

14.3.2 Supplies

The supplies we need for this activity are a long wire, cables, a dc power supply, a compass, a bucket, paperclips, nail, and alligators (see figure 14.9).

Figure 14.9. From left to right: cables, power supply, a compass, a bucket, and alligators.

14.3.3 Identifying the poles of a compass

We recall that opposite poles of a magnet attract each other, and the same poles repel each other. Using Earth's magnetic field, identify the north and the south poles of the compass shown in figure 14.10.

Figure 14.10. The compass

The north pole (red or blue?)
The south pole (red or blue?)
Note: the geographic north is Earth's magnetic south pole, and the geographic south is Earth's magnetic north pole (see figure 14.11).

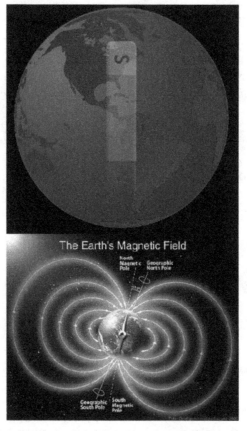

Figure 14.11. Earth's geographic and magnetic poles. Courtesy of Peter Reid, The University of Edinburgh.

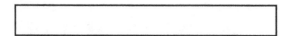

Figure 14.12. The magnetic field directions of a bar magnet.

Mapping the magnetic field of a bar magnet

1. Place the bar magnet at the center of a clean sheet of paper on the table. Hold the compass about a centimeter above the paper close to one of the bar magnet's ends. Draw a small arrow under the compass on the paper that points from the south to the compass's north pole.

2. Now move the compass and repeat the procedure in step 1 until we return where we started. In this manner, map out the magnetic field throughout the region of the piece of paper.

3. Identify and label the north and south poles of the bar magnet on the paper. Sketch what we have on the paper with the north and south pole identified in the space provided.

14.3.4 Mapping the magnetic field of a current in a long straight wire

1. *Making predictions*: suppose there are two long current-carrying wires perpendicular to this page. The direction of the currents in the first wire is out of the page, and in the second wire is into the page (see figure 14.13). Using the right-hand rule I, find the direction of the magnetic field produced. Show the direction by tracing a curved arrow (a clockwise or counter-

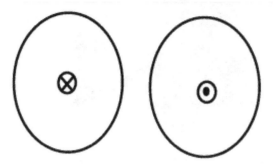

Figure 14.13. Magnetic field of a long wire carrying a current into the page and out of the page.

clockwise) on the space provided (figure 14.13).

   Note: the symbol X in a small circle indicates the current is directed into the page and a dot in a small circle indicates the current is directed out of the page.

2. *Experimental verification*: next, we will verify our prediction for a long current-carrying wire's magnetic field direction. We follow the procedure listed below:

   (a) Straighten out your longest wire as best as you can and feed the wire through the small hole in the cardboard bucket (see figure 14.14).

Figure 14.14. The long wire through the cardboard bucket.

Figure 14.15. The wire connected to the wooden block.

(b) Connect the DC power supply to the wooden block with red and black cables. Also, attach the two ends of the wire under the terminals on the wooden block (make sure there is electrical contact, see figure 14.15).

(c) Make sure that the wire is extended and vertical on both sides of the bucket.

(d) Turn on the power supply. Increase the voltage by turning the knob to about 30%–60%. There is a circuit breaker button on the front/back of the supply. If the power goes off (which it does at high voltage), turn off the power (turn the power knob down first) and press this button.
Caution: the wire could get hot at high voltage!

(e) In figure 14.16, the big circle is the edge of the bucket, and the small circle is the wire as you view it from the top. Put the appropriate sign to show the wire's current is moving into or out of the bucket. Following a similar procedure used for the bar magnet, move the compass in a circle centered about the wire and find the magnetic field's direction. The direction of the magnetic field would be the direction of the north pole of the compass. Show the direction of the magnetic field on the big circle in figure 14.16 (use a clockwise or counterclockwise arrow).

Does the experimental magnetic field agree with our prediction?
If your answer is no, figure out what went wrong.

(f) Turn the power off. Switch the polarity of the power supply cable.

(g) Turn the power back on and repeat procedure (e). (This time use figure 14.17 to show the directions.)

Does the experimental magnetic field agree with our prediction?

If your answer is no, figure out what went wrong.

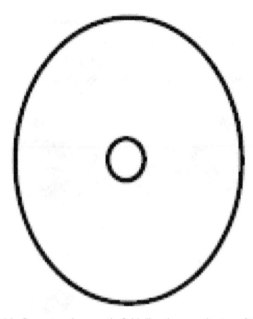

Figure 14.16. Current and magnetic field directions on the top of the bucket.

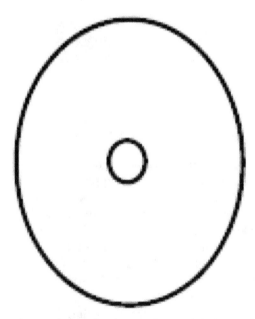

Figure 14.17. Current and magnetic field directions on the top of the bucket (for switched polarities).

14.3.5 Mapping the magnetic field of a current in a solenoid

Next, we shall study the magnetic field produced by a current in a solenoid (see figure 14.18). Like the previous section, we shall make predictions for the magnetic field's directions and verify our predictions experimentally.

Figure 14.18. A solenoid.

1. Place the solenoid near the center of a white printing paper (see figure 14.19).

Figure 14.19. A solenoid placed near the center of a printing paper.

2. Using cables, connect the solenoid's red lead to the red lead and the black lead to the black lead of the power supply (see figure 14.20).
3. For the circuit we just made (figure 14.20), the current in the solenoid flows from left to right (i.e., red lead to black lead of the solenoid). In the wire turning in a circle in the solenoid, this current is directed in a counterclockwise direction as we look from the solenoid's right side (i.e., the black lead side). For this current flow, using right-hand rule III (refer to figure 14.3 in section 14.1), identify the north and south pole of the solenoid and sketch the magnetic field directions in figure 14.21.
4. Turn on the power supply. Increase the voltage by turning the knob to about 30%–60%. Hold the compass close to one of the ends of the solenoid (black or red leads). Trace a small arrow on the paper showing the direction of the magnetic field.

   Note: the direction of the arrow is the direction of the north pole of the compass.

14-15

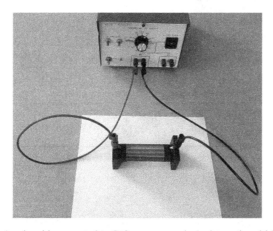

Figure 14.20. A solenoid connected to DC power supply (red to red and black to black).

Figure 14.21. The sketch for the magnetic field direction predicted.

5. Now move the compass and repeat the procedure in step 4 until we return where we started. In this manner, map out the magnetic field throughout the region on the paper and identify the solenoid's north and south poles.

6. Sketch what we have on the paper in figure 14.22. (In the sketch, identify the north and south poles and the direction of the current.) Does the experimental magnetic field agree with our prediction?

Figure 14.22. The sketch for the magnetic field direction experimentally determined.

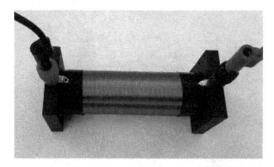

Figure 14.23. The solenoid with switched polarity.

If your answer is no, figure out what went wrong.

7. Turn the power off. Switch the polarity of the solenoid (see figure 14.23). Turn the power back on and repeat steps 4–6 but use figure 14.23 for your sketch. (Another white paper is recommended.)

8. Turn the power supply off. Leave the solenoid connected to the power supply.

14.3.6 Ferromagnetism and the electromagnet

1. Put a paper clip inside the solenoid and turn the power back on.

2. Now turn the solenoid from a horizontal position to a vertical position slowly and lift it off the table.

Turn the power off. What happens to the paper clip?

3. Explain what we have observed when the power supply is on and off.

4. Turn the power back on and try to pick the paper clip with one of the solenoid ends. Did we succeed? Yes or no.

5. Put the nail inside the solenoid and try to pick up the paper clip with the nail's tip. What do we see that is different from what we saw in step 4?

6. The nail has some ferromagnetic material in it. Explain why the new solenoid-magnet is so much stronger than the direct solenoid.

   What you have built is a small electromagnet.

7. Remove the nail from the solenoid and use it to try to pick up the paper clip with its sharp end. Does it work? If it does, explain why?

8. Hold the head of the nail and whack the other end on the ground several times. Repeat this for the head of the nail too.

9. Try to pick up the paper clip with the nail. If it does pick the paper clip, keep on whacking it on the ground until we fail to pick it up. Explain what happened.

14.3.7 Result and conclusion

1. How do we find the direction of the magnetic force acting on a current-carrying wire due to an external magnetic field? The direction of the external magnetic field in the region and the direction of the current are known.

2. Write a statement of the effect of having a ferromagnetic material in the external magnetic field.

3. Briefly explain the concept of an electromagnet.

4. Write a brief overview of what we accomplished and concluded in this activity.

IOP Publishing

Virtual and Real Labs for Introductory Physics II

Optics, modern physics, and electromagnetism

Daniel Erenso

Chapter 15

Faraday's law

In the last chapter, we studied the magnetic field properties produced by a bar magnet, a long current-carrying wire, and a solenoid. One way or another, these magnetic fields result from some charge flow (an electric current) that we can explain on a micro- or macro-scale. This chapter introduces us to the reverse process of how an electric current is produced in a closed conducting wire using a magnetic field. We begin with a brief introduction to motional electromotive force (emf), magnetic flux, Faraday's law and Lenz's law. We then do virtual and real labs to understand the relations among changing magnetic flux, Faraday's law and Lenz's law. In the virtual lab, we use the PhTH simulation involving a simple bar magnet, a coil, a bulb and an ammeter. In the real lab, we use a bar magnet, a solenoid, a resistor and an oscilloscope to study an induced voltage (induced emf) and the direction of the resulting electric current in the solenoid. In both the real and the virtual labs, we use an approach based on prior predictions of the quantities followed by experimental verification.

15.1 Basic theory

15.1.1 Motional electromotive force

Consider a rectangular conducting wire connected to a resistor of resistance, R, is pulled to the right with a constant speed v (see figure 15.1). The loop is placed in a uniform magnetic field B directed into the page. The width of the rectangular wire is h. This results in is an induced motional electromotive force (induced voltage) in the loop ΔV_{in} given by

$$\Delta V_{in} = vBh. \tag{15.1}$$

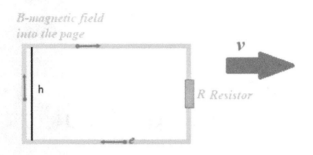

Figure 15.1. Motional electromotive force (emf).

The magnitude of the induced current I_{in} is

$$I_{in} = \frac{\Delta V_{in}}{R} = \frac{vBh}{R}. \tag{15.2}$$

15.1.2 Magnetic flux, Faraday's law and Lenz's law

We saw that a motional emf could be induced in a closed conducting loop moving inside a uniform magnetic field. Such emf generally depends on the change of the magnetic flux through the area bounded by the conducting loop and is given by

$$\varepsilon = -\frac{\Delta \Phi}{\Delta t}, \tag{15.3}$$

where

$$\Phi = BA \cos(\theta) \tag{15.4}$$

is the magnetic flux through the area A bounded by the conducting loop, and θ is the angle between the normal to the area and the magnetic field $\vec{B}$.

Lenz's law determines the direction of the induced current resulting from the induced voltage. It states the induced current must have a direction that gives an induced magnetic field, which leads to an induced magnetic flux that opposes the cause for the change in the flux. That means if the cause for the induced current is due to a decrease in magnetic flux, the induced current must produce an induced magnetic field that increases the magnetic flux.

15.2 Virtual lab: *Faraday's law*

15.2.1 Introduction

The objectives of this virtual lab are
- To study how magnetic flux changes and the effect of such changes in a closed conducting wire.
- To have a deeper understanding of Faraday's law.
- To see what causes induction of voltage in closed conducting loops.
- To practice how we determine the direction of the induced magnetic field and current.

To this end, to open the simulation window in figure 15.2 go to/click on https:// phet.colorado.edu/en/simulation/faradays-law. Click play, leading us to the window for this lab shown in figure 15.3.

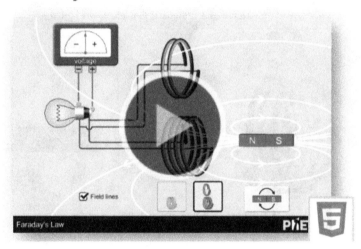

Figure 15.2. Faraday's law simulation.

15.2.2 Part I: a bar magnet—*north pole*

In this part, we will study the voltage-induced properties when we insert and pull out a bar magnet in a coil of conducting wire connected to a voltmeter and a light bulb.
1. Move the magnet as slow and far away as possible from the coil. Line up the magnet (the north pole) with the axis of the coil (see figure 15.4).

2. Click the box 'Field lines'. How many field lines pass through the coil when the magnet is far from the coil?

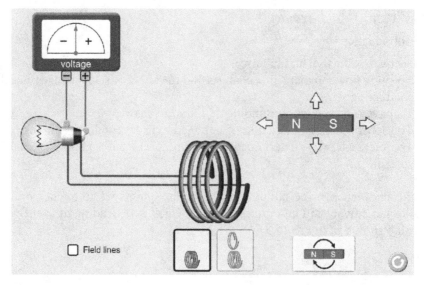

Figure 15.3. Faraday's law lab.

Figure 15.4. Lining up the magnet with the axis of the coil.

3. Observe the light's brightness from the bulb connected to the coil and the reading on the voltmeter hooked across the bulb on the top left side. At this point, we read zero voltage and see no light. Now fully insert the magnet, leave it inside for a few seconds, maybe for 5–10 s, and then pull the magnet out. When we insert and pull out the magnet, consider three speeds: slow, medium, and fast. As much as possible, try to keep the center of the magnet coinciding with the coil's center. Please repeat this step as many as possible such that we are confident to answer the following questions.

   (a) How many field lines pass through the coil when the magnet is fully inserted?

(b) If you were sitting at the place where the funny looking guy (the Alien) is sitting in figure 15.5, when you insert the north pole of the magnet, is the magnetic field directed into or out of the coil?

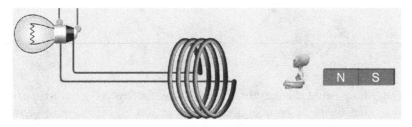

Figure 15.5. A guy sitting on the right side and looking in the direction of the magnetic field.

(c) When you were inserting the magnet's north pole, is the magnetic flux increasing or decreasing? Explain why? (*Remember what you studied in the previous activity about the magnetic field strength and direction for a bar magnet*).

(d) Using the appropriate symbols, show these magnetic field directions (lines) before and after inserting the magnet in figure 15.6. (Note: use fewer symbols for fewer field lines.)

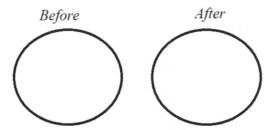

Figure 15.6. Magnetic field direction before and after inserting the bar magnet.

(e) Based on your answer in part (c), what must be the direction of *the induced magnetic field* (if you were to view it from the same position in figure 15.5)? Show the direction using the appropriate symbol in figure 15.7.

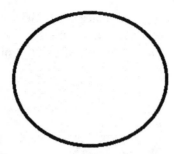

Figure 15.7. Induced magnetic field.

(f) Such induced magnetic field is created by an induced current in a clockwise or counterclockwise direction (if you were to view it from the same position in figure 15.5)?

(g) When you were inserting the north pole of the magnet, did you read a positive or negative voltage? What does that tell us about the direction of the current? Is this current in clockwise or counterclockwise direction (if you view it from the same position in figure 15.5)? Is this current direction consistent with what we predicted in (f)?

(h) When you were pulling the magnet out, is the magnetic flux increasing or decreasing?

(i) Based on your answer in part (h), what must be the direction of *the induced magnetic field*. Show the direction using the appropriate symbol below (figure 15.8).

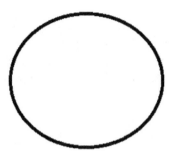

Figure 15.8. Induced magnetic field.

(j) What is the direction of the induced current that gives this induced magnetic field? Is it in a clockwise or counterclockwise direction (if you were to view it from the same position in figure 15.5)?

(k) When you were pulling the magnet out, what did you read for the voltage? Positive or negative voltage? What does that tell us about the direction of the current? Is this current in a clockwise or counter-clockwise direction (if you were to view it from the same position shown in figure 15.5)? Is this current consistent with what we predicted in (j)?

(l) You did insert and pull out the magnet at different speeds (slow, medium, and fast). Briefly discuss what happened to the voltage induced, the brightness of the light from the bulb, and its relationship to how fast we insert the magnet.

15.2.3 Part II: a bar magnet—*south pole*

Next, we are interested in answering the same questions presented in Part I. However, this time we insert the south pole of the magnet. To this end, flip the bar magnet's polarity such that the south pole is pointing at the coil.

1. Move the magnet as slow and as far away as possible from the coil. Line up the magnet (the south pole) with the coil's axis (see figure 15.9).

Figure 15.9. The south pole of the bar magnet pointing at the coil.

2. How many field lines pass through the coil?

3. Observe the voltmeter and the brightness of the light from the bulb. Now fully insert the magnet, leave it inside for a few seconds, and then pull the magnet out. When we insert in and pull out the magnet, consider three speeds: slow, medium, and fast. Keep the center of the magnet coinciding with the center of the coil.
   (a) How many field lines pass through the coil when we fully insert the magnet?

   (b) If you were sitting at the place where the funny looking guy (the Alien) is sitting in figure 15.10, when you insert the south pole of the magnet, is the magnetic field directed into or out of the coil?

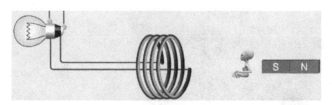

Figure 15.10. A guy sitting on the right side and looking in the direction of the magnetic field.

(c) When you were inserting the magnet's south pole, is the magnetic flux increasing or decreasing? Explain why?

(d) Using the appropriate symbols, show these magnetic field directions (lines) before and after insertion of the bar magnet in figure 15.11.

Before   *After*

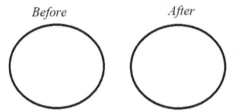

Figure 15.11. Magnetic field direction before and after the insertion of the magnet.

(e) Based on your answer in part (c), what must be the direction of *the induced magnetic field* (viewed from the same position in figure 15.10)? Show the direction using the appropriate symbol in figure 15.12.

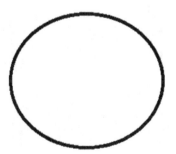

Figure 15.12. Induced magnetic field.

(f) Is the direction of the induced current clockwise or counterclockwise (viewed from the same position in figure 15.10)?

(g) When you were inserting the magnet in, what did you read for the voltage? Positive or negative voltage? What does that tell us about the direction of the current? Is this current in a clockwise or counter-clockwise direction (viewed from the same position in figure 15.10)? Is this current consistent with what we predicted in (f)?

(h) When you were pulling the magnet out, is the magnetic flux increasing or decreasing?

(i) Based on your answer in part (h), what must be the direction of *the induced magnetic field.* Show the direction using the appropriate symbol in figure 15.13.

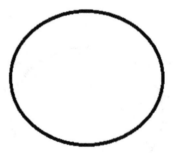

Figure 15.13. The direction of the induced magnetic field.

(j) What is the direction of the induced current (viewed from the same position in figure 15.10)?

(k) When you were pulling the magnet out what did you read for the voltage? Positive or negative voltage? What does that tell you about the direction of the current? Is this current in clockwise or counterclockwise direction (viewed from the same position in figure 15.10)? Is this current consistent with what you predicted in (j)?

(l) Briefly discuss what happened to the voltage induced, the brightness of the light from the bulb, and its relationship to how fast we insert the magnet.

15.2.4 Result and conclusion

1. Briefly discuss the similarities and differences that we observed in Part I and Part II.

2. When the bar magnet is sitting inside the coil, please explain why we read zero voltage and saw no light in both Part I and Part II of this lab.

3. Write a brief overview of what we accomplished and concluded in this activity.

15.3 Real lab: *Faraday's law*

15.3.1 Objective

The objectives of this lab are
- To have a better understanding of magnetic flux and change in magnetic flux.
- To make a quantitative and qualitative experimental observation of Faraday's law, induced voltage, and induced current.
- To learn how to determine the direction of an induced current.

15.3.2 Supplies

The supplies we need for this activity are cables, power supply, a compass, a bar magnet, an oscilloscope, a 47 Ω resistor, BNC cables, and alligators.

15.3.3 Procedure

Finding the poles of the magnet
Identify the poles of the compass in figure 15.14.

The north pole (red or blue?)

The south pole (red or blue?)

Figure 15.14. The compass.

15.3.4 Current flow in a solenoid and magnetic poles

1. Place the solenoid on a white paper and connect it to the dc power supply (see figure 15.15). Turn on the power supply and set the power to about 30%–60%.
2. Following the direction of the current flow from the power supply, identify the current flow in the solenoid (whether it is from left to right or from right to left). Label this direction using a big arrow on the white paper and also in figure 15.15.
3. Using the compass, identify the north and south poles for the solenoid in figure 15.15. Label these poles on the white paper and also in figure 15.15.

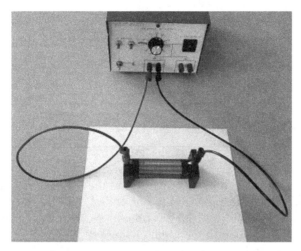

Figure 15.15. Connecting the solenoid to the power supply.

4. When we look from the solenoid's black lead side, is the current in the circular wires wound around on the solenoid flowing in a clockwise or counterclockwise direction? (Right-hand rule III!)

5. Place the solenoid on another white paper, switch the polarities (see figure 15.16). Turn the power switch off when we do this part.
   Turn the power back on and make sure the voltage is set to about 30%–60%.
6. Repeat steps 2 and 3. Use figure 15.16 for your answers.

Figure 15.16. The solenoid with switched polarity.

7. When we look from the solenoid's black lead side, is the current flowing in a clockwise or counterclockwise direction?

8. Turn the power off and place the solenoid on another white paper. Switch back the polarities for the solenoid (i.e. figure 15.15).
9. Disconnect the cables from the power supply.
   Note: keep the two papers, where you labeled the polarities and the poles, for reference.

15.3.5 Inducing voltage (current) by a magnet—*north pole*

1. Turn the oscilloscope and set the ground at the center of the screen (the *x*-axis).
2. Set the vertical sensitivity to 100 mV and the horizontal sweep time to 1 s.
3. Connect the solenoid's red lead to one end of the 47Ω resistor and the black lead to the other end. (Use the breadboard and regular red and black cables.)
4. Connect a BNC cable to the input channel of the oscilloscope. Connect the red lead of the BNC cable to the red side and the black lead to the black side. (*The circuit you build resembles figure* 15.17.)

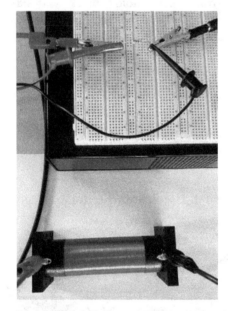

Figure 15.17. A resistor hooked to a solenoid and an oscilloscope.

5. Bring the north pole of the magnet close to the solenoid's black lead side and hold it still with the magnet lined up with the solenoid's axis for about 3–5 s.
6. Now insert the magnet fully, fast, and gently. Keep the fully inserted magnet stationary for about 5 s and pull it out about the same way inserted. We see something that resembles figure 15.18. Press stop to freeze the image and take a picture, and this will be one image that we will analyze.

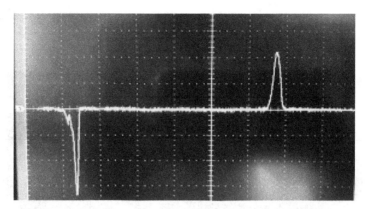

Figure 15.18. Induced voltage: insert in, keep stationary (for 5 s) and pull out (north pole).

7. Fully insert in and immediately pull the magnet out of the solenoid (keep the north pole directed in the same way). You see a trace resembling figure 15.19. Press stop to freeze the image and take a snapshot, and this is a second image that we will analyze.

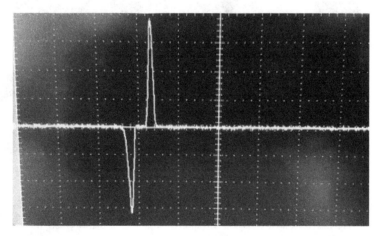

Figure 15.19. Induced voltage: insert in and pull out immediately (north pole).

8. To better understand and the straightforward analyses of the images we took above, repeat what we have done for the following cases.
   (a) Different insertion and pulling speeds.
   (b) Different duration for keeping the magnet stationary inside the solenoid.

   Please make a note of the changes we observe.

Inducing voltage (current) by a magnet—south pole
Repeat steps 5–8 for the south pole.

1. The traces (for the south pole) on the oscilloscope for step 6 resembles figure 15.20 and for step 7 resembles figure 15.21

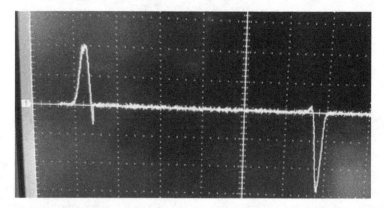

Figure 15.20. Induced voltage: insert in, keep stationary (for 5 s) and pull out (south pole).

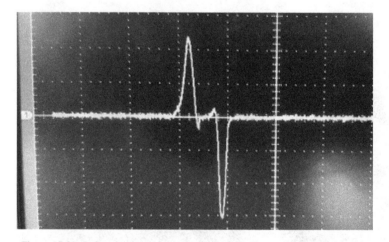

Figure 15.21. Induced voltage: insert in and pull out immediately (south pole).

15.3.6 Data analyses (for the north pole)

A. North pole image one (your image for figure 15.18)

1. What is the amplitude of the positive and negative voltage induced both in divisions and volts?

2. For the positive voltage, what is the direction of the induced current in the solenoid? *You may answer from red to black lead of the solenoid or black to red.*

3. For the negative voltage, what is the direction of the induced current in the solenoid?
   You may answer like you did in step 2.

4. When we insert the magnet's north pole, is the magnetic field directed into or out of the solenoid?

5. We have noticed that the negative voltage is a consequence of inserting the magnet into the solenoid. Was the flux increasing or decreasing when we were inserting the magnet?

6. Based on our answers to questions 4 and 5, what should be the direction of the induced magnetic field? Should it be in the same or opposite direction to the external magnetic field (i.e., the magnetic field by the bar magnet)?

7. According to our answer to question 6, which end of the solenoid must be the north pole (black or red lead)?

8. Are the answers to questions 3 and 7 consistent with our results determined for current directions and magnetic poles?

9. The positive voltage is a result of pulling out the magnet from the solenoid. Was the flux increasing or decreasing when we were pulling out the magnet?

10. What should be the direction of the induced magnetic field? Should it be in the same or opposite direction to the external magnetic field (i.e., the magnetic field by the bar magnet)?

11. Which end of the solenoid must be the north pole (black or red lead)?

12. Are the answers to questions 2 and 11 consistent with our results determined for current directions and magnetic poles?

13. The magnitude of the positive voltage and negative voltage may not be the same. Explain why? If we get the same, also explain why. There is sound physical reasoning in either case.

14. For how long the flux in the solenoid has been constant in the interval between we inserted in and pulled out the magnet.
    In DIV:

    In seconds:

B. North pole image two (your image for figure 15.19)
  1. What are the amplitudes of the positive and negative voltage induced? Positive voltage amplitude in DIV and volts:

     Negative voltage amplitude in DIV and volts:

2. For how long has the flux in the solenoid been constant in the interval between we inserted in and pulled out the magnet.

In DIV:

In seconds:

3. Explain the significant difference between the results demonstrated in image one and image two.

15.3.7 Data analyses (for the south pole)

A. South pole image one (your image for figure 15.20)

1. What are the amplitudes of the positive and negative voltage induced?

Positive voltage amplitude in DIV and volts:

Negative voltage amplitude in DIV and volts:

2. For the positive voltage, the induced current in the solenoid is directed from what to what?

3. For the negative voltage, the induced current in the solenoid is directed from what to what?

4. When we insert the magnet's south pole, the magnetic field is directed into or out of the solenoid?

5. This time, the negative voltage is a result of pulling the magnet out of the solenoid. Was the flux increasing or decreasing when we were pulling out the magnet?

6. Based on our answers to questions 4 and 5, what should be the direction of the induced magnetic field? Should it be in the same or opposite direction to the external magnetic field (i.e., the magnetic field by the bar magnet)?

7. According to our answer to question 6, which end of the solenoid must be the north pole (black or red lead)?

8. Are the answers to questions 3 and 7 consistent with our results determined for current directions and magnetic poles for the solenoid?

9. For the south pole, the positive voltage is a result of inserting the magnet into the solenoid. Was the flux increasing or decreasing when we were inserting the south pole of the magnet?

10. What should be the direction of the induced magnetic field? Should it be in the same or opposite direction to the external magnetic field?

11. Which end of the solenoid must be the north pole (black or red lead)?

12. Are the answers to questions 2 and 11 consistent with our results determined for current directions and magnetic poles of the solenoid?

13. The magnitude of the positive voltage and negative voltage may not be the same. Explain why we get different values? If we get the same, also explain why. There is sound physical reasoning in either case.

14. For how long the flux in the solenoid has been constant in the interval between we inserted in and pulled out the magnet.
    In DIV:

    In seconds:

B. South pole image two (your image for figure 15.21)
  1. What is the amplitude of the positive and negative voltage induced?
     Positive voltage amplitude in DIV and volts:

     Negative voltage amplitude in DIV and volts:

  2. For how long has the flux in the solenoid been constant in the interval between we inserted in and pulled out the magnet.
     In DIV:

     In seconds:

  3. Explain the significant difference between the results demonstrated in image one and image two?

15.3.8 Result and conclusion

Write a brief overview of what we accomplished and concluded in the virtual and real labs.

IOP Publishing

Virtual and Real Labs for Introductory Physics II
Optics, modern physics, and electromagnetism
Daniel Erenso

Chapter 16

Induction and RL circuits

As we have studied in the last chapter, when the magnetic field inside a coil of wire (or a solenoid) changes, the magnetic flux changes resulting in an induced current opposing the flux change. This induced current affects the current in an electrical circuit when it has such a coil of wire, known as an inductor. This chapter studies an electrical circuit with an inductor and a resistor (RL circuit). We begin with a brief introduction to an inductor's properties and a qualitative and quantitative description of an RL circuit when a direct current is on and off. We then perform virtual and real labs. In the virtual lab, using the PhTH simulation, we build an RL circuit with an on and off switch to determine the circuit's inductive time constant by analyzing the inductor's voltage vs time. This study is replicated in the real lab using an oscilloscope.

16.1 Basic theory

An inductor: a coil of conducting wire

Magnetic flux in an inductor: when an inductor is part of an electrical circuit, the current passing through the inductor produces a magnetic field. The flux through the coil due to this magnetic field is given by

$$\Phi_B = LI, \tag{16.1}$$

where I is the current and L is called the inductance of the coil. The inductance depends on the geometrical properties of the coil and is measured in units of Henry (H),

$$1 \text{ Henry} = 1 \text{ T} \cdot \text{m}^2/\text{A} \tag{16.2}$$

Faraday's law: change in magnetic flux results in an induced voltage,

$$|\Delta V_{\text{in}}| = \left| \frac{\Delta \Phi_B}{\Delta t} \right|. \tag{16.3}$$

Induced voltage in an inductor: if the current changes with time, so does the magnetic field. Such change results in the change of the magnetic flux in the inductor that leads to an induced voltage,

$$|\Delta V_{\text{in}}| = L \left| \frac{\Delta I}{\Delta t} \right|, \tag{16.4}$$

where ΔI is the change in current over a time interval Δt.

An RL circuit: in an RL dc circuit, an inductor with inductance L and a resistor with resistance R are connected to a dc power supply, ΔV_s. In figure 16.1, when the

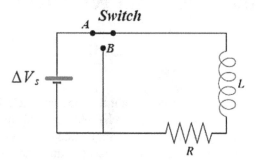

Figure 16.1. RL circuit connected to a dc power supply.

power is on (switch is at A), the inductor acts as an open circuit at the initial time ($t = 0$) and a short circuit after a long time (at $t = \infty$). The current in the circuit is given by

$$I(t) = I_{\text{max}}(1 - e^{-t/\tau}), \tag{16.5}$$

where τ is the inductive time constant that depends on the inductance L and the equivalent resistance of the circuit R_{eq}

$$\tau = \frac{L}{R_{\text{eq}}}, \tag{16.6}$$

and I_{max} is the maximum current in the circuit

$$I_{\text{max}} = \frac{\Delta V_s}{R_{\text{eq}}}. \tag{16.7}$$

If we disconnect the power supply by flipping the switch to B, there will be an induced current flowing in a reverse direction due to the inductor's sudden magnetic flux change. This current is given by

$$I(t') = I_0' e^{-t'/\tau'}, \tag{16.8}$$

where I_0' is the current in the circuit when we flip the switch from A to B, and τ' is the inductive time constant.

Half-life time (1/2): the amount of time for the voltage across the inductor reaches one-half of its limiting value, and it is related to the inductive time constant by

$$t_{1/2} = \tau \ln 2. \tag{16.9}$$

We will use this relation to determine the inductive time constant (figure 16.2).

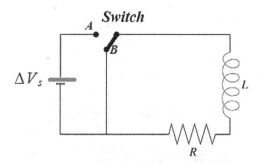

Figure 16.2. RL circuit disconnected from a dc power supply (switch at B).

16.2 Virtual lab: *RL circuit*

16.2.1 Introduction

The objectives of this virtual lab are

- To construct and study a dc circuit's properties with a resistor and an inductor (dc RL circuit).
- To study the properties of an inductor in a dc circuit.
- To have a deeper understanding of Faraday's law.
- To experimentally measure the inductive time constant in an RL circuit when an inductor is being 'charged' and then 'discharged'.

To this end, go to/click onhttps://phet.colorado.edu/en/simulation/legacy/circuit-construction-kit-ac to open the simulation kit shown in figure 16.3.

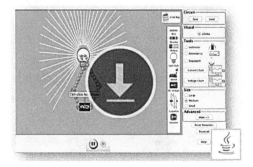

Figure 16.3. Circuit construction kit (AC + DC), Virtual Lab

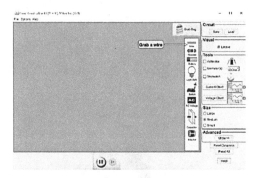

Figure 16.4. Opened circuit construction kit.

Download and open the file, and it leads us to the simulation window shown in figure 16.4.

16.2.2 Part I: 'charging' an inductor

1. Using the dc voltage (battery), two switches, two resistors, an inductor, and wires, construct the circuit shown in figure 16.5.

Figure 16.5. Dc RL circuit.

2. Set the resistance $R_1 = R_2 = R = 10\Omega$, the inductance $L = 10H$, and the source voltage, $\Delta V_s = 5V$.
3. Click on the 'voltage chart' and hook up the red and black leads across the inductor. Once again, click on the 'voltage chart' and hook up the red and black leads across the second resistor. Click on 'stopwatch' (see figure 16.6).

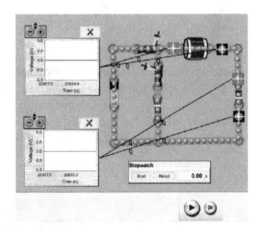

Figure 16.6. Measuring the voltage across the inductor and resistor in a dc RL circuit.

4. Keep both switches off, and if accidentally turned it on and current is flowing in the circuit, turned it back off and 'discharge' the inductor.
5. *Setting up 'the voltage charts'*: as we can see in figure 16.7, the vertical axis has four divisions, and the horizontal axis has two divisions. To set this up for the vertical axis, we can click on '−' or '+'. On the other hand, to set the time axis, click on play and stop just before the dotted vertical line reaches the voltage axis. Then use the slow play to line the dotted vertical line with the voltage axis.

If we have carried all the steps above successfully, then we can proceed to the next step.

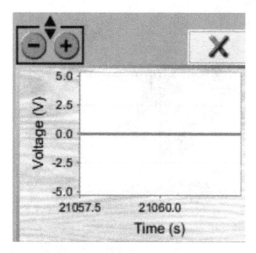

Figure 16.7. The setting for the voltage chart.

6. What is the voltage per division (V/div) for the vertical axis and time per division (t/div) for the horizontal axis?

7. Now turn the source switch on (the one on the horizontal line in figure 16.6) and click play. We see the voltage across the inductor $\Delta V_L(t)$ and the voltage across the resistor $\Delta V_R(t)$ being displayed as a function of time t. When we stop the simulation, for example, after about 2.5 s, we see the traces for $\Delta V_L(t)$ and $\Delta V_R(t)$ that look like the graphs shown in figure 16.8.

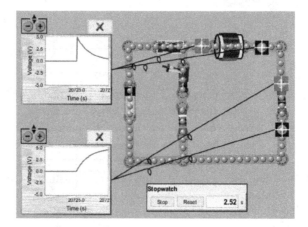

Figure 16.8. The voltage across the inductor and the resistor after the switch is turned on ($t \simeq 2.5s$).

8. Use the fast and slow play buttons till you see full traces for $\Delta V_L(t)$ and $\Delta V_R(t)$ (for example, see figure 16.9).

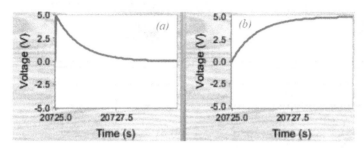

Figure 16.9. The voltage across the inductor (a) and the resistor (b) as a function of time.

9. Figure 16.9(a) and (b) show that, at the initial time, $\Delta V_L(t)$ is equal to the source voltage ($\Delta V_L = \Delta V_s = 5$ V) and $\Delta V_R(t)$ is zero ($\Delta V_R = 0$). However, as time goes on $\Delta V_L(t)$ goes to zero, and $\Delta V_R(t)$ approaches the source voltage. Give a brief explanation in terms of Faraday's law why this is happening. What does this generally tell us about inductors' property in a dc RL circuit at $t = 0$ and $t = \infty$?

10. At any given point of time after the switch is turned on, what relation can we establish between the source voltage ΔV_s and the total voltage across the inductor and the resistor ($\Delta V_L(t) + \Delta V_R(t)$)?

11. Using the full display for $\Delta V_L(t)$ that you see on the screen, find the values for the quantities listed in table 16.1.

12. Find the theoretical and experimental values for the inductive time constant in seconds and record the values in table 16.2.

Table 16.1. Measured voltage across the inductor.

Voltages	In DIV	
Maximum voltage across the inductor		
Half of the maximum voltage		
Times	**In DIV**	**In seconds**
$t_{1/2}$: the time at which the voltage reduces to half of the maximum voltage		

Table 16.2. Inductive time constant.

Theoretical inductive time constant, $\tau_{theo} = \dfrac{L}{R}$	
Experimental inductive time constant, $\tau_{exp} = \dfrac{t_{1/2}}{\ln 2}$	

13. What is the equivalent resistance, R_{eq} of the circuit?

14. Using the experimental value for the inductive time constant τ and the equivalent resistance R_{eq} find the inductance for the inductor.

15. Find the percent difference between the theoretical and experimental values for the inductance. Do the results for the experimental values agree with the theoretical, within the experimental uncertainties? If not, explain why?

16.2.3 Part II: 'discharging' the inductor

At this stage, we should have the simulation window that looks like what we see in figure 16.10.

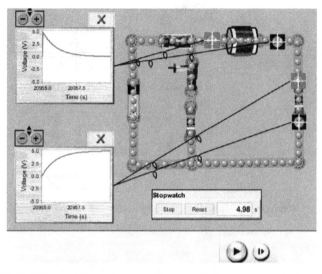

Figure 16.10. What you should have on your screen before you move to the next part.

1. Please do not play the simulation until we are told to do so. Turn the first switch off (what does this do to the source voltage?) and then turn on the second switch. Find the equivalent resistance of the circuit R_{eq}?

2. Reset the stopwatch. Move the black lead of the voltage chart from the first resistor to the top end of the second resistor to measure the total voltage across the two resistors (see figure 16.11).

3. Run the simulation (play) and stop it after about 3 s. We must see what we see in figure 16.12.

4. Adjust the voltage axis for both charts until we can see the full traces for the voltage (see figure 16.13).

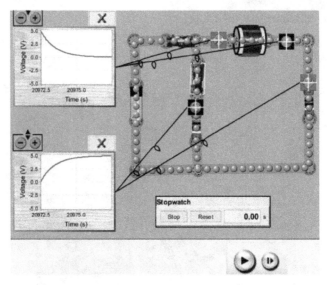

Figure 16.11. Measuring the voltage across the two resistors.

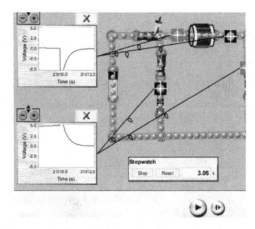

Figure 16.12. The simulation after it runs after about 3 s.

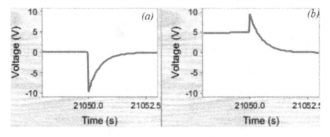

Figure 16.13. The complete trace for the voltage across the inductor (a) and the resistor (b).

5. What is the voltage per division (V/ div) for the vertical axis?

6. In Part I, we found the voltage across the inductor $\Delta V_L(t)$ starts from positive ($\Delta V_L \simeq 5$ V) and decreases to zero. However this time it starts from negative $\Delta V_L' \simeq -9.5$ V and approaches zero. Explain (a) why the voltage is negative and (b) why $\Delta V_L' \simeq 2\,|\Delta V_L|$. (c) Does $\Delta V_L' \simeq 2\,|\Delta V_L|$ imply a violation of conservation of energy?

7. Use fast and slow play to get full display for $\Delta V_L'(t)$ and $\Delta V_R'(t)$ (see figures 16.14(a) and (b)).

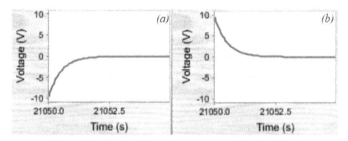

Figure 16.14. The full trace for $\Delta V_L'(t)$ (a) and $\Delta V_R'(t)$ (b).

8. What relation can we establish between $\Delta V_L'(t)$ and $\Delta V_R'(t)$, where $\Delta V_R'(t)$ is the total voltage across the two resistors.

9. Using the full display for $\Delta V_L'(t)$, find the values for the quantities listed in table 16.3.

Table 16.3. The measured voltages across the inductor.

Voltages	In DIV	
Maximum voltage across the inductor		
Half of the maximum voltage		
Times	**In DIV**	**In seconds**
$t_{1/2}$: the time at which the voltage reduces to half of the maximum voltage		

10. Find the theoretical and experimental values for the inductive time constant in seconds and record the values in table 16.4.

Table 16.4. Inductive time constant.

Theoretical inductive time constant, $\tau_{\text{theo}} = \dfrac{L}{R_{\text{eq}}}$	
Experimental inductive time constant, $\tau_{\text{exp}} = \dfrac{t_{1/2}}{\ln 2}$	

11. Using the experimental value for the inductive time constant, find the inductance L.

12. Do the results for the experimental values agree with the theoretical, within the experimental uncertainties? If not, explain why?

16.2.4 Result and conclusion

Write a brief overview of what we accomplished and concluded in this activity.

16.3 Real lab: *RL circuits*

16.3.1 Objective

The objectives of this lab are
- To build an RL circuit.
- To learn more about the use of a TTL signal as a switch.
- To experimentally determine the inductive time constant in an RL circuit.

16.3.2 Supplies

In this activity, we need alligators, cables, and an inductor (10 mH), a resistor (3.33 kΩ), an oscilloscope, a function generator, and BNC cables (figure 16.15). Measure the resistance of the resistor and the inductor and record the values.

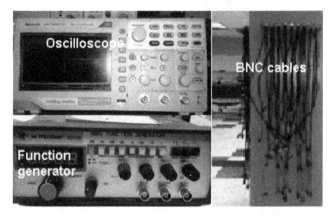

Figure 16.15. The oscilloscope, function generator, and BNC cables.

16.3.3 Building the RL circuit

Using the resistor, the inductor, and the function generator, we will build the RL circuit shown in figure 16.16.

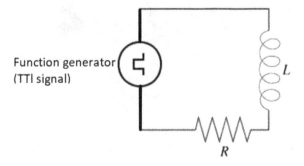

Figure 16.16. An RL circuit.

16-13

1. Connect the resistor and the capacitor in series on the breadboard.
2. Connect a BNC cable to the function generator on the TTL/COMS output jack.
3. Connect the red lead of the BNC cable to the resistor end that is not connected to the inductor.
4. Connect the BNC cable's black lead to the inductor end that is not connected to the resistor.
5. Turn on the function generator. Push the button for 10 kHz. Get a frequency reading of about 5 kHz by adjusting the COARSE and FINE knobs.
6. Set the TTL signal's amplitude to the maximum value by turning the AMPL(itude) knob of the function generator.
7. Turn the oscilloscope on and set the ground (for the channel that you want to use, CH1 or CH2) at the time axis (the middle of the screen, e.g., figure 16.17).

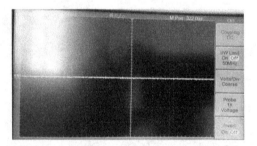

Figure 16.17. Ground set at the middle of the screen.

8. Connect another BNC cable to the oscilloscope you are using.
9. Connect the red lead of the BNC cable to the resistor end that is not connected to the inductor.
10. Connect the BNC cable's black lead to the inductor end that is not connected to the resistor.
11. We should see the TTL signal that resembles the trace shown in figure 16.18.

Figure 16.18. TTL Signal.

16.3.4 Analyzing the voltage across an inductor

1. Connect the oscilloscope leads so that the scope reads the voltage across the inductor ΔV_L. We should see the voltage across the inductor that looks like figure 16.19.

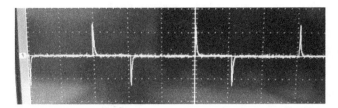

Figure 16.19. The voltage across the inductor.

2. Move the ground to the bottom of the screen (see, as an example, figure 16.20).

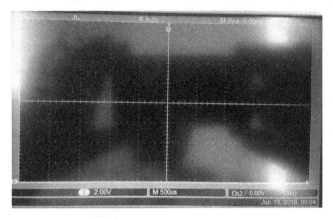

Figure 16.20. The new ground set at the bottom.

3. Adjust the horizontal sweep time and the vertical sensitivity knobs until we see an enlarged trace for the positive voltage. Move this trace horizontally using the knob that controls the horizontal POSITION till we see something that resembles the trace in figure 16.21.

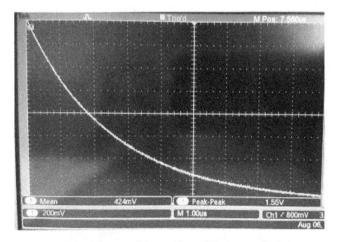

Figure 16.21. The trace of the positive voltage across the inductor.

4. Record the horizontal sweep time setting in sec/div.

5. For the enlarged trace of the positive voltage we see on the screen, find the values for the quantities listed in table 16.5.

Table 16.5. Measured voltage and time for the inductor.

Voltages	In DIV	
Maximum voltage across the inductor (on the screen that you see)		
Half of the maximum voltage		
Times	**In DIV**	**In seconds**
$t_{1/2}$: the time at which the voltage reduces to half of the maximum voltage		

6. Find the theoretical and experimental values for the inductive time constant in seconds and record the values in table 16.6.

Table 16.6. Inductive time constant.

Theoretical inductive time constant, $\tau_{\text{theo}} = \frac{L}{R}$	
Experimental inductive time constant, $\tau_{\text{exp}} = \frac{t_{1/2}}{\ln 2}$	

7. Using the experimental value for the inductive time constant and the relation

$$\tau_{\text{exp}} = \frac{L}{R},\qquad(16.10)$$

find the inductance L.

16.3.5 Result and conclusion

Write a brief overview of what we accomplished and concluded in this activity

IOP Publishing

Virtual and Real Labs for Introductory Physics II

Optics, modern physics, and electromagnetism

Daniel Erenso

Chapter 17

Introduction to ac circuits

All the circuits we studied up to this point have a direct current (dc) source. This chapter introduces electrical circuits with an alternating current (ac) source voltage. In particular, we consider electrical circuits with a sinusoidal source voltage and a resistor, a capacitor, or an inductor. The potential difference across these circuit elements depends on the current, and a quantity is known as impedance. We first present a brief introduction to the theory predicting the impedance for a resistor, a capacitor, and an inductor connected to a sinusoidal ac source voltage. We then carry out a PhTH simulation to determine the impedance for a resistor, a capacitor, and an inductor by connecting each of these elements to a sinusoidal source voltage. In the real part of the activity, using an ammeter that measures the root-mean-square current (I) and an oscilloscope that displays the ac voltage in real-time, we re-analyze the impedance of a resistor, a capacitor, and an inductor in an ac circuit. In both the simulation and real labs, we explore the frequency dependence of each of these circuit elements' impedance.

17.1 Basic theory

Sinusoidal voltage signal:

$$\Delta V(t) = \Delta V_{\max} \sin(\omega t), \tag{17.1}$$

where $\Delta V_{\max}$ is the maximum voltage (the amplitude), ω is the angular frequency related to the linear frequency f by

$$\omega = 2\pi f. \tag{17.2}$$

Root-mean-square voltage, or rms voltage :

$$\Delta V_{\mathrm{rms}} = \frac{\Delta V_{\max}}{\sqrt{2}}. \tag{17.3}$$

doi:10.1088/978-0-7503-3715-1ch17

Root-mean-square current, or rms current :

$$I_{\text{rms}} = \frac{I_{\text{max}}}{\sqrt{2}}.$$ (17.4)

Complex number, z: a complex number is defined as

$$z = a + jb,$$ (17.5)

where

$$a = \text{Real part}$$
$$b = \text{Imaginary part}$$ (17.6)
$$j = \sqrt{-1}.$$

Complex conjugate, z^*: complex conjugate of the complex number z

$$z^* = a - jb.$$ (17.7)

Magnitude of a complex number, $|z|$:

$$|z| = \sqrt{zz^*} = \sqrt{a^2 + b^2}.$$ (17.8)

Resistive impedance, Z_R:

$$Z_R = R \Rightarrow |Z_R| = R.$$ (17.9)

Capacitive impedance, Z_C:

$$Z_C = \frac{1}{j\omega C} = -\frac{j}{\omega C} \Rightarrow |Z_C| = \frac{1}{\omega C}.$$ (17.10)

Inductive impedance, Z_L:

$$Z_L = j\omega L \Rightarrow |Z_L| = \omega L.$$ (17.11)

17.2 Virtual lab: *ac circuits and impedance*

17.2.1 Introduction

The objectives of this virtual lab are
- To get introduced to ac circuits and impedances for a resistor, capacitor, and inductor in an ac circuit.
- To study the voltage across a resistor, a capacitor, or an inductor when the source voltage is an ac source.

We will be using the same simulation kit we used in the previous lab. To open this simulation kit go to/click on https://phet.colorado.edu/en/simulation/legacy/circuit-construction-kit-ac. You should see what you saw in the previous virtual lab shown in figure 17.1. Download and open the file, and it leads us to the simulation window shown in figure 17.2.

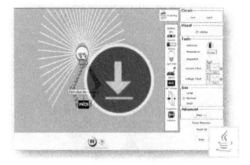

Figure 17.1. Circuit construction kit (AC + DC), Virtual Lab.

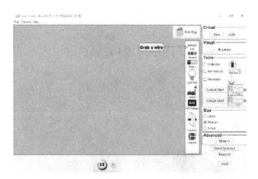

Figure 17.2. Opened circuit construction kit.

17.2.2 Part I: ac circuit with a capacitor

1. Using the ac voltage, a switch, a capacitor, and wires, construct the circuit shown in figure 17.3.

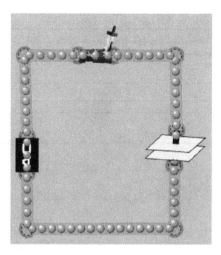

Figure 17.3. Ac circuit with a capacitor.

2. Set the capacitance, $C = 0.05F$. Right-click on the ac voltage and select 'change frequency', and set the frequency, $f = 0.5$ Hz.
3. Click on the 'voltage chart' and hook up the red and black leads across the capacitor.
4. Click on the 'current chart' and hook up the lead anywhere on the electric current path. Since the switch is open, you see zero current and zero voltage (see figure 17.4).

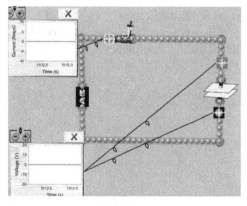

Figure 17.4. An ac circuit with a capacitor with the addition of the 'current chart' and 'voltage chart'.

5. *Theoretical*: find the *angular frequency* of the ac voltage, ω, and the capacitor's impedance for the capacitor, $|Z_{the}|$. Record the values at the appropriate place in table 17.1.

Table 17.1. Theoretical and experimental values for the capacitor.

Capacitor	At 0.5 Hz frequency	At 2 Hz frequency
ω in (rad/s)		
$|Z_{the}|$ in (Ω)		
ΔV_{max} in (V)		
ΔV_{rms} in (V)		
I_{rms} in (A)		
$|Z_{exp}|$ in (Ω)		

6. *Experimental*: now, turn the switch on and wait until we see a clear sinusoidal trace for both the voltage and the current. Then read the maximum voltage ΔV_{max} and the maximum current I_{max}. Calculate the *rms voltage* ΔV_{rms} and *rms current* I_{rms}. Using the *rms voltage* and *rms current* determine the experimental impedance $|Z_{exp}|$ for the capacitor. Record the values at the appropriate places in table 17.1. We can zoom in or zoom out 'the voltage chart' and 'the current chart' by clicking the plus or the minus signs on the chart for better reading.
7. Turn the switch off, discharge the capacitor and change the source voltage frequency to $f = 2.0$ Hz.
8. Turn the switch back on and repeat steps 5 and 6.
9. Based on the results in table 17.1, discuss how the capacitor's impedance changes with the ac voltage frequency.

17.2.3 Part II: ac circuit with an inductor

1. Turn the switch off, remove the capacitor from the circuit and replace it with an inductor (see figure 17.5). Set the inductance to $L = 10$ H. Reset the frequency of the ac voltage to $f = 0.5$ Hz.

Figure 17.5. Ac circuit with an inductor.

2. Keep the 'voltage chart' and 'current chart'. This time, the 'voltage chart' reads the voltage across the inductor.
3. *Theoretical*: find the *angular frequency* ω and the *magnitude of the impedance* for the capacitor $|Z_{the}|$. Record the values at the appropriate place in table 17.2.

Table 17.2. Theoretical and experimental values for the inductor.

Inductor	At 0.5 Hz frequency	At 2 Hz frequency
ω in (rad/s)		
$|Z_{the}|$ in (Ω)		
ΔV_{max} in (V)		
ΔV_{rms} in (V)		
I_{rms} in (A)		
$|Z_{exp}|$ in (Ω)		

4. *Experimental*: turn the switch back on and read the maximum voltage ΔV_{max} and the maximum current I_{max}. Calculate the *rms voltage* ΔV_{rms} and *rms current* I_{rms} across the capacitor. Using ΔV_{rms} and I_{rms} determine the experimental impedance for the inductor $|Z_{exp}|$. Record the values at the appropriate places in table 17.2.
5. Turn the switch off and change the frequency of the source voltage to $f = 2.0$ Hz.

6. Discharge the inductor and turn the switch back on and repeat steps 3 and 4.
7. Based on the results on table 17.2, discuss how the inductor's impedance changes with the ac voltage frequency.

17.2.4 Part II: ac circuit with a resistor

1. Turn the switch off, remove the inductor from the circuit and replace it with a resistor (see figure 17.6). Set the resistance to $R = 100\ \Omega$. Reset the frequency of the ac voltage to $f = 0.5$ Hz.

Figure 17.6. Ac circuit with a resistor.

2. Keep the 'voltage chart' and 'current chart'. This time, the 'voltage chart' reads the voltage across the resistor.
3. *Theoretical*: find the *angular frequency* of the ac voltage, ω and the *magnitude of the impedance* $|Z_{\text{the}}|$. Record the values at the appropriate place in table 17.3.
4. *Experimental*: now turn the switch back on, read ΔV_{max}, I_{max} and calculate ΔV_{rms} and I_{rms}. Using ΔV_{rms} and I_{rms} determine the experimental impedance $|Z_{\text{exp}}|$. Record the values at the appropriate places in table 17.3.
5. Turn the switch off and change the frequency to $f = 2.0$ Hz. Then turn the switch back on and repeat steps 3 and 4.
6. Based on the results in table 17.3, discuss how the resistor's impedance changes with the ac voltage frequency.

Table 17.3. Theoretical and experimental values for the resistor.

Resistor	At 0.5 Hz frequency	At 2 Hz frequency
ω in (rad/s)		
$|Z_{\text{the}}|$ in (Ω)		
ΔV_{max} in (V)		
ΔV_{rms} in (V)		
I_{rms} in (A)		
$|Z_{\text{exp}}|$ in (Ω)		

17.2.5 Result and conclusion

Write a brief overview of what we accomplished and concluded in this activity.

17.3 Real lab: *ac circuits and impedance*

17.3.1 Objectives

The objectives of this lab are

- To study voltage and current in an ac circuit.
- To get introduced to root-mean-square (rms) current and voltage in ac circuit.
- To experimentally and theoretically study impedances for a resistor, a capacitor, and an inductor.

17.3.2 Supplies

The supplies we need are alligators, cables, an oscilloscope, a function generator, an ammeter, BNC cables (figure 16.5). We also need a resistor ($R = 100\ \Omega$), a capacitor ($C = 1.0\ \mu$ F), and an inductor ($L = 10$ mH). Measure the resistance for each and record the values.

17.3.3 AC circuit with a resistor

In this section, we will study the impedance for a resistor in an ac circuit. For a resistor with resistance, R, the magnitude of the impedance, Z_R, theoretically, is determined from the resistance

$$Z_R = R \Rightarrow |Z_R| = R. \tag{17.12}$$

The experimental value, on the other hand, is calculated from the rms voltage (ΔV_{rms}) and the rms current (I_{rms}) using the relation

$$|Z_R| = \frac{\Delta V_{rms}}{I_{rms}}, \tag{17.13}$$

where

$$\Delta V_{rms} = \frac{\Delta V_{max}}{\sqrt{2}},\ I_{rms} = \frac{I_{max}}{\sqrt{2}}. \tag{17.14}$$

Thus to find the experimental value for the impedance we need to measure the maximum voltage (ΔV_{max}) and current (I_{max}). To this end, we build the circuit shown in figure 17.7 on the breadboard. Try to build this circuit just from the diagram. What we need is the resistor, the ammeter, and the sinusoidal wave from the function generator. If we succeeded, we could proceed to step 5.

1. Connect the **output** for the function generator with the **BNC cable.**
2. Connect the black lead of the function generator BNC cable to one end of the resistor on the breadboard.
3. Using a regular cable, connect the other end of the resistor to the black lead of the ammeter.
4. Using a regular cable, connect the red lead of the ammeter to the red lead of the BNC cable. This step completes the circuit.

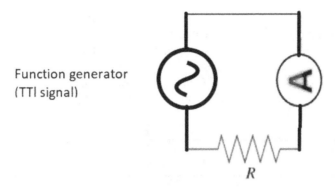

Function generator
(TTI signal)

R

Figure 17.7. The circuit diagram.

5. Hook another BNC cable across the resistor (red lead to one end and black lead to the other end of the resistor) and connect it to CH 1 or CH 2 input of the oscilloscope.
6. Set the ammeter to ac by flipping the black switch to ac.
7. Turn both the oscilloscope and the signal generator on. Like the previous activities, set the ground to the middle of the screen. Set the vertical sensitivity knob to $5V/DIV$.
8. Turn the function generator and select the sine wave. Set the AMPL(itude) to the maximum value and the FREQUENCY to 60 Hz.
9. Play with the horizontal sweep time knob and the other knobs and buttons that we know about the oscilloscope until we see a precise sinusoidal wave (see figure 17.8).

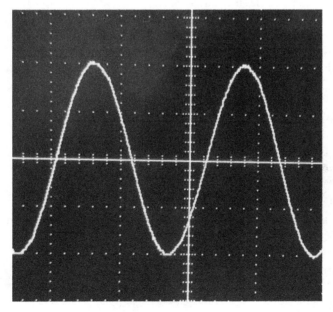

Figure 17.8. A sine wave.

10. If everything is done as instructed; we must see that the ammeter needle has moved and read some positive current. What we read is the rms current (I_{rms}).
11. Using the linear frequency f, we had set for the function generator, calculate the angular frequency, $\omega = 2\pi f$. Record the value in the space provided in

Table 17.4. Theoretical and experimental values for the resistor.

Resistor	At 60 Hz frequency	At 600 Hz frequency
$\omega = 2\pi f$ in (rad/s)		
$|Z_{the}|$ in (Ω)		
ΔV_{max} in (V)		
ΔV_{rms} in (V)		
I_{rms} in (A)		
$|Z_{exp}|$ in (Ω)		

table 17.4.
12. Calculate the theoretical value for the impedance, $|Z_{the}|$. Record the value in the space provided in table 17.4.
13. Read the maximum voltage (ΔV_{max}) from the oscilloscope and calculate the corresponding rms voltage (ΔV_{rms}). Record the values in the space provided in table 17.4.
14. Read rms current (I_{rms}) from the ammeter and record the value in the space provided in table 17.4.
15. Estimate the uncertainties for the voltage and current.
    Voltage uncertainty:
    Current uncertainty:
16. Calculate the fractional uncertainties for the voltage and current we measured.
    FU voltage:
    FU current:
17. Calculate the experimental impedance, $|Z_{exp}|$, along with the uncertainty and record the values in the space provided in table 17.4.
18. Does the experimental value agree with the theoretical value within the experimental uncertainties?

19. Change the frequency of the function generator to 600 Hz and repeat the procedures 11–14, 17, and 18. (Use the same fractional uncertainties.)
20. For the 600 Hz frequency, does the experimental value agree with the theoretical value within the experimental uncertainties?

21. From our results for the two frequencies, what can we conclude about the relationship between the frequency of the ac voltage and the resistor's impedance?

17.3.4 AC circuit with a capacitor

In this section, we will study the impedance of a capacitor. For a capacitor with capacitance, C, the impedance, Z_C, is given by

$$Z_C = -\frac{j}{\omega C} \Rightarrow |Z_C| = \frac{1}{\omega C}. \qquad (17.15)$$

Experimentally, the impedance is determined using the same relation we used for the resistor

$$|Z_C| = \frac{\Delta V_{rms}}{I_{rms}}.$$

We follow a similar procedure to the previous section to find the experimental rms voltage and current.

1. Turn the function generator off.
2. While keeping everything else the same, replace the resistor with the capacitor.
3. Turn the function generator back on and set the FREQUENCY to 100 Hz.
4. Play with the horizontal sweep time and vertical sensitivity knobs until we see a precise sine wave (see figure 17.8).
5. If everything is done as instructed, we read some positive current on the ammeter.
6. Repeat steps 11–14 that we used for the resistor. (But this time record the values in table 17.5.)
7. Estimate the uncertainties for the measured voltage and current.
   Voltage uncertainty:
   Current uncertainty:
8. Calculate the fractional uncertainties.
   FU voltage:
   FU current:

Table 17.5. Theoretical and experimental values for the capacitor.

Capacitor	At 100 Hz frequency	At 200 Hz frequency
$\omega = 2\pi f$ in (rad/s)		
$|Z_{the}|$ in (Ω)		
ΔV_{max} in (V)		
ΔV_{rms} in (V)		
I_{rms} in (A)		
$|Z_{exp}|$ in (Ω)		

9. Calculate the experimental impedance and the uncertainty and record the values in the space provided in table 17.5.
10. Does the experimental value is close enough to the theoretical value within the experimental uncertainties?

11. Change the function generator's frequency to 200 Hz and repeat steps 4–6 and 9.
12. For the 200 Hz frequency, is our experimental value close enough to the theoretical value within the experimental uncertainties?

13. From our results for the two frequencies, what can we conclude about the relationship between the frequency of the ac voltage and the capacitor's impedance?

17.3.5 AC circuit with an inductor

For an inductor with inductance, L, the theoretical impedance, Z_L, is determined using

$$Z_L = j\omega L \Rightarrow |Z_L| = \omega L. \tag{17.16}$$

Experimentally, we also use the relation

$$|Z_L| = \frac{\Delta V_{rms}}{I_{rms}}. \qquad (17.17)$$

In this activity, the rms current is directly read from the ammeter. The rms voltage is calculated by reading the peak voltage from the oscilloscope first.
1. Turn the function generator off.
2. While keeping everything else the same, replace the capacitor with the inductor.
3. Turn the function generator back on and set the FREQUENCY to 1000 Hz.
4. If everything is done as instructed, we read some positive current on the ammeter.
5. Repeat procedures 11–14 we used for the resistor. (But this time record the values on table 17.6.)

Table 17.6. Theoretical and experimental values for the inductor.

Inductor	At 1000 Hz frequency	At 6000 Hz frequency
$\omega = 2\pi f$ in (rad/s)		
$\|Z_{the}\|$ in (Ω)		
ΔV_{max} in (V)		
ΔV_{rms} in (V)		
I_{rms} in (A)		
$\|Z_{exp}\|$ in (Ω)		

6. Estimate the uncertainties for the voltage and current you measured.
   Voltage uncertainty:
   Current uncertainty:
7. Calculate the fractional uncertainties.
   FU voltage:
   FU current:
8. Calculate the experimental impedance for the inductor and the uncertainty. Record the value in the space provided in table 17.6.
9. Is the experimental value close enough to the theoretical value within the experimental uncertainties?
10. Change the frequency of the function generator to 6000 Hz and repeat procedures 4–5 and 8.
11. For the 2000 Hz frequency, is the experimental value close enough to the theoretical value within the experimental uncertainties?

12. From our results for the two frequencies, what can we conclude about the relationship between the frequency of the ac voltage and the inductor's impedance?

17.3.6 Result and conclusion

Write a brief overview of what we accomplished and concluded in this activity.

IOP Publishing

Virtual and Real Labs for Introductory Physics II
Optics, modern physics, and electromagnetism
Daniel Erenso

Chapter 18

AC filters and resonance

In the previous chapter, we studied electrical impedance in a resistor, a capacitor, and an inductor when connected to an ac source voltage separately. This chapter introduces how a resistor and a capacitor serve as an ac filter in an electrical circuit with an ac source voltage. After briefly discussing the theory associated with ac filter circuits that includes RC (resistor and capacitor) and RLC (resistor, inductor, and capacitor) circuits, we perform two virtual and two real labs. In the first part of the virtual lab, using the PhTH simulation, we study how a resistor serves as a high-pass and a capacitor as a low-pass filter by analyzing the output and input voltages. We then carry out a similar analysis in an RLC circuit to study resonance in the virtual lab's second part. We then replicate our studies in two real labs; one for an ac RC circuit and another for an ac RLC circuit. We also analyze a radio signal used as an ac source voltage in an RC circuit in the first part of the real lab. We use a two-channel oscilloscope to measure the input and output voltages and a function generator to produce a sinusoidal ac voltage with a variable frequency.

18.1 Basic theory

Equivalent impedance: the good thing about the equivalent impedance is that we can determine it like resistors. For parallel connections

$$\frac{1}{Z_{eq}} = \frac{1}{Z_1} + \frac{1}{Z_2} \tag{18.1}$$

and series combination

$$Z_{eq} = Z_1 + Z_2. \tag{18.2}$$

doi:10.1088/978-0-7503-3715-1ch18   18-1

Gain (G): the gain across a resistor, a capacitor, or an inductor is defined as

$$G = \frac{|\Delta V_{\text{out}}|}{|\Delta V_{\text{in}}|} = \left| \frac{\Delta V_{\text{out}}}{\Delta V_{\text{in}}} \right|. \tag{18.3}$$

Note that

$$\text{when } G \text{ is } \begin{cases} < 1 & \text{the signal is attenuated} \\ = 1 & \text{the signal is passed} \\ > 1 & \text{the signal is amplified} \end{cases}. \tag{18.4}$$

High-pass filter: in an RC ac circuit, when the output voltage is taken across the resistor

$$G = \frac{|\Delta V_{\text{out}}|}{|\Delta V_{\text{in}}|} = \frac{|I(t)Z_R|}{|I(t)Z_{\text{total}}|} = \frac{|Z_R|}{|Z_{\text{total}}|} = \frac{R}{\sqrt{R^2 + \left(\dfrac{1}{\omega C}\right)^2}}. \tag{18.5}$$

For high-pass filter

$$\text{when } \omega \begin{cases} \to \infty, & G = 1 \text{ (passed)} \\ \to 0, & G = 0 \text{ (attenuated)} \end{cases}. \tag{18.6}$$

Low-pass filter: in an RC ac circuit, when the output voltage is taken across the capacitor

$$G = \frac{|\Delta V_{\text{out}}|}{|\Delta V_{\text{in}}|} = \frac{|I(t)Z_C|}{|I(t)Z_{\text{total}}|} = \frac{|Z_C|}{|Z_{\text{total}}|} = \frac{\dfrac{1}{\omega C}}{\sqrt{R^2 + \left(\dfrac{1}{\omega C}\right)^2}}. \tag{18.7}$$

For law-pass filter

$$\text{when } \omega \begin{cases} \to 0, & G = 1 \text{ (passed)} \\ \to \infty, & G = 0 \text{ (attenuated)} \end{cases}. \tag{18.8}$$

Transition frequency f_{tr}: in an ac circuit the total impedance has two terms: one is frequency dependent and the other is not. As we vary the frequency the magnitude of these two terms can become equal. This frequency is called the transition frequency. For the RC ac circuit, the total impedance is

$$Z_{\text{total}} = R + j\left(-\frac{1}{\omega C}\right), \tag{18.9}$$

here the frequency independent term is R and the frequency dependent term is $-1/\omega C$. For the transition frequency ω_{tr}, we have

$$R = \frac{1}{\omega_{tr}C} \Rightarrow \omega_{tr} = \frac{1}{RC} \Rightarrow f_{tr} = \frac{1}{2\pi RC}. \tag{18.10}$$

Resonance and resonance frequency f_{res}: in series an RLC ac circuit with the output voltage taken across the capacitor, the gain is given by

$$G = \frac{|\Delta V_{out}|}{|\Delta V_{in}|} = \frac{|I(t)Z_C|}{|I(t)Z_{total}|} = \frac{|Z_C|}{|Z_{total}|} = \frac{\frac{1}{\omega C}}{\sqrt{R^2 + \left(\omega L - \frac{1}{\omega C}\right)^2}}, \tag{18.11}$$

which gives

$$\begin{cases} \omega \rightarrow 0, & G = 1 \text{ (passed)} \\ \omega \rightarrow \infty & G = 0 \text{ (attenuated)} \end{cases}. \tag{18.12}$$

However, at a particular frequency, f_{res}, the gain takes a maximum value and becomes greater than one. This is known as resonance and this frequency is called resonance frequency. It is given by

$$f_{res} = \frac{1}{2\pi}\sqrt{\frac{2L - R^2C}{2L^2C}} = \frac{1}{2\pi}\sqrt{\frac{1}{LC} - \frac{R^2}{2L^2}}. \tag{18.13}$$

For a negligibly small resistance, $R \simeq 0$,

$$f_{res} = \frac{1}{2\pi\sqrt{LC}}. \tag{18.14}$$

18.2 Virtual lab I: *ac filters*

18.2.1 Introduction

The objectives of this virtual lab are
- To study an RC circuit with an ac source voltage.
- To measure input voltage, output voltage, and study gain.
- To study high-pass and low-pass filters in an RC ac circuit.

We will be using the same simulation kit we used in the previous lab. To open this simulation kit go to/click on https://phet.colorado.edu/en/simulation/legacy/circuit-construction-kit-ac and you see the same window you saw in the previous lab, shown in figure 18.1. Download and open the file, and it takes us to the simulation window shown in figure 18.2.

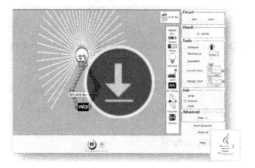

Figure 18.1. Circuit construction kit (AC + DC).

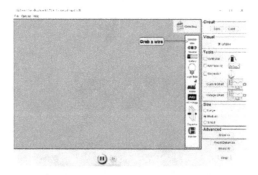

Figure 18.2. Opened circuit construction kit.

18.2.2 Part I: high-pass filter

1. Using the ac voltage, a switch, a capacitor, a resistor, and wires, construct the circuit shown in figure 18.3.

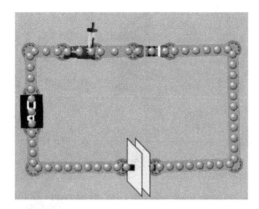

Figure 18.3. Ac RC circuit.

2. Set the capacitance, $C = 0.2F$, and the resistance $R = 5\,\Omega$. Using the values for R and C, calculate the transition frequency

$$f_{tr} = \frac{1}{2\pi RC} \tag{18.15}$$

and the theoretical value for the gain G_{theo} across the resistor

$$G_{theo} = \frac{R}{\sqrt{R^2 + \left(\dfrac{1}{2\pi f_{tr}\, C}\right)^2}}.$$

3. Click on the 'voltage chart' and hook up the red and black leads across the ac source voltage.
4. Click one more time on the 'voltage chart' and hook up the red and black leads across the resistor (see figure 18.4).
5. When we calculated the transition frequency, we must have gotten, $f_{tr} \simeq 0.16\ \text{Hz}$. Now set the input voltage (the ac source) frequency to the transition frequency by $f_{tr} \simeq 0.16\text{Hz}$. Also, make sure the input voltage amplitude is 10 V (i.e., $\Delta V_{in,max} = 10\ \text{V}$).

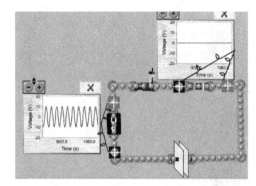

Figure 18.4. 'Voltage chart' reading across the ac source and the resistor.

6. Turn the switch on and read the amplitude of the output voltage (the voltage across the resistor, $\Delta V_{out,max} = \Delta V_{R,max}$) with the uncertainties and calculate the corresponding fractional uncertainty.

7. Find the experimental value for the gain G_{exp} across the resistor

$$G_{exp} = \frac{|\Delta V_{out,max}|}{|\Delta V_{in,max}|}, \tag{18.16}$$

with the uncertainties.

What we do next is pretty much the same as what we have done for the transition frequency. The only difference is that we will take similar measurements for several frequencies to verify that a resistor is indeed a high-pass filter by doing a graphical analysis. So we need to read and record the values for the input and output voltage ($\Delta V_{out,max}$) for different frequencies listed in table 18.1.

8. Read $\Delta V_{out,max}$ and $\Delta V_{in,max}$ for each of the frequencies listed in table 18.1 and calculate the gain, G. Record the values in the appropriate places in table 18.1.

9. Using Excel, make a graph for the gain (G) vs frequency (f) and sketch the graph in figure 18.5.

10. Calculate the value for $\log_{10}(100 \times f)$ and record the values in the space provided in table 18.1.

Table 18.1. Measured frequency, input and output voltage.

Frequency (Hz)	$\log_{10}(100 \times f)$	$\Delta V_{in,max}(V)$	$\Delta V_{out,max}\ (V)$	$G = \dfrac{\|\Delta V_{out,\ max}\|}{\|\Delta V_{in,\ max}\|}$
0.01				
0.02				
0.03				
0.04				
0.07				
0.09				
0.15				
0.2				
0.3				
0.7				
1.2				
1.7				
1.9				
2.0				

11. Using Excel, make a graph for the gain (G) vs $\log_{10}(100 \times f)$ and sketch the graph in figure 18.6.

18.2.3 Part II: low-pass filter

1. We now proceed to verify a capacitor is a low-pass filter. However, first, we will predict the behavior of the gain as a function of the frequency. To this end, we recall that theoretically, the gain for a capacitor is given by

$$G_{theo} = \frac{\dfrac{1}{\omega C}}{\sqrt{R^2 + \left(\dfrac{1}{\omega C}\right)^2}} = \frac{\dfrac{1}{2\pi f C}}{\sqrt{R^2 + \left(\dfrac{1}{2\pi f C}\right)^2}}. \tag{18.17}$$

Graph title:_____

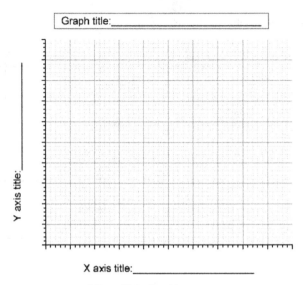

Y axis title:_____

X axis title:_____

Figure 18.5. Graphing page.

Graph title:_____

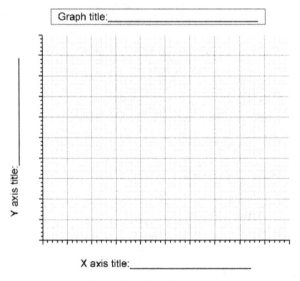

Y axis title:_____

X axis title:_____

Figure 18.6. Graphing page.

Using the given values for the capacitance, $C = 0.2F$ and the resistance $R = 5\,\Omega$., find the theoretical value for the gain (G_{theo}) for the two frequencies in table 18.2 and record the values in the space provided in this table.

2. Using what we have learned in graphing G vs $\log_{10}(100 \times f)$ for the resistor in Part I and the values we obtained for G_{theo} and $\log_{10}(100 \times f)$, make a sketch that predicts G vs $\log_{10}(100 \times f)$ for the capacitor in figure 18.7.

Table 18.2. Predicted gain values for the capacitor.

Frequency (Hz)	$\log_{10}(100 \times f)$	$\Delta V_{in,max}$ (V)	$\Delta V_{out,max}$ (V)	G_{theo}	G_{exp}
0.01					
2.00					

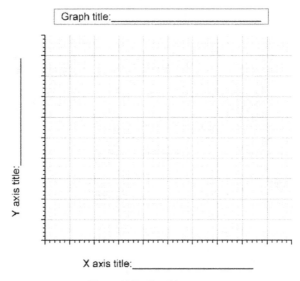

Figure 18.7. Graphing page.

3. Now, turn the switch off and disconnect the 'voltage chart' from the resistor and connect it across the capacitor.
4. Turn the switch back on and for the two frequencies listed in table 18.2, read the voltage amplitudes for $\Delta V_{in,max}$ and $\Delta V_{out,max}$ and record the values in the space provided in the same table.
5. Calculate the experimental values for the gain, G_{exp}.
6. Does the experimental values agree with our predictions within the experimental uncertainties? If not, explain why?

18.2.4 Result and conclusion

Write a brief overview of what we accomplished and concluded in this activity.

18.3 Virtual lab II: *RLC circuit and resonance*

18.3.1 Introduction

The objectives of this virtual lab are
- To learn about an ac RLC circuit.
- To study more about input voltage, output voltage, and gain using an RLC circuit.
- To get introduced to the concept of resonance.

We will be using for one last time the same simulation kit we used in the previous lab. To open this simulation kit again go to/click on https://phet.colorado.edu/en/simulation/legacy/circuit-construction-kit-ac and we see the window shown in figure 18.8. Download and open the file to display the simulation window shown in figure 18.9.

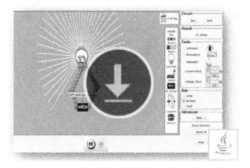

Figure 18.8. Circuit construction kit (AC + DC).

Figure 18.9. Opened circuit construction kit.

18.3.2 RLC circuit

1. Using the ac voltage, a switch, a resistor, an inductor, a capacitor, and wires, construct the circuit shown in figure 18.10.

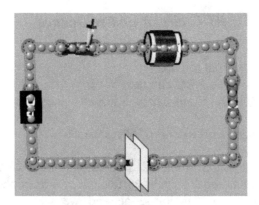

Figure 18.10. Ac RLC circuit.

2. Set the resistance $R = 2.5\,\Omega$, the inductance $L = 10H$, the capacitance, $C = 0.2F$. Using these values, calculate the resonance frequency,

$$f_{\text{res}} = \frac{1}{2\pi}\sqrt{\frac{2L - R^2C}{2L^2C}}.\qquad(18.18)$$

3. Click on the 'voltage chart' and hook up the red and black leads across the ac source voltage.
4. Click one more time on the 'voltage chart' and hook up the red and black leads across the capacitor (see figure 18.11).

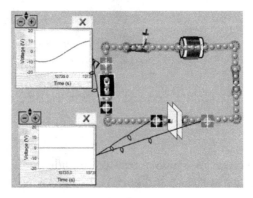

Figure 18.11. RLC circuit input and output voltage (across the capacitor) measurement.

5. What we will do next is pretty much the same as what we did in the previous lab. We need to read and record the values for the input voltage ($\Delta V_{in,max}$) and the corresponding output voltage ($\Delta V_{out,max}$) for the different frequencies listed in table 18.3. We then calculate the gain, G

$$G = \frac{|\Delta V_{out,\,max}|}{|\Delta V_{in,\,max}|}.$$ (18.19)

Table 18.3. Measured frequency, input and output voltage.

Frequency (Hz)	$\Delta V_{in,max}$ (V)	$\Delta V_{out,max}$ (V)	G
0.01			
0.02			
0.03			
0.04			
0.07			
0.06			
0.07			
0.08			
0.09			
0.10			
0.11			
0.12			
0.13			
0.16			
0.2			
0.25			
0.3			
0.4			
0.9			
1.5			

After we changed the frequency, we wait a minute or two before reading for the output voltage across the capacitor. This time allows us to have an accurate trace for the output voltage.

18.3.3 The resonance gain curve

1. Using Excel, make a graph for the gain (G) vs frequency (f). We should see a range of frequencies where the gain increases to a peak and decreases. The frequency at which we see the peak is the resonance frequency.

2. *Experimental resonance frequency*: determine the resonance frequency from the graph.

3. Sketch the graph in figure 18.12.

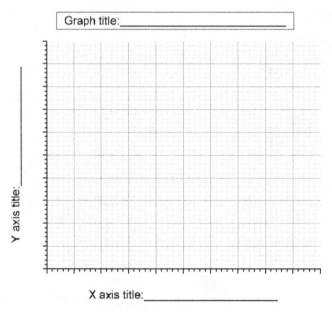

Figure 18.12. Graphing page.

4. Find the percent difference between the theoretical and experimental resonant frequencies. Does the result confirm the experimental values agree with our prediction within the experimental uncertainties? If not, explain why?

18.3.4 Result and conclusion

Write a brief overview of what we accomplished and concluded in this activity.

18.4 Real lab I: ac *filters*

18.4.1 Objectives

The objectives of this lab are
- To study an RC circuit with an ac source voltage.
- To measure input voltage, output voltage, and study gain.
- To study high-pass and low-pass filters in an RC ac circuit.

18.4.2 Supplies

The supplies we need are alligators, cables, an oscilloscope, a function generator, BNC cables, headphones, a small radio, a resistor ($R = 68\Omega$) and a capacitor ($C = 1.0\mu F = 10^{-6}F$). Measure the resistance for the resistor and the capacitor and record the values.

18.4.3 The transition frequency

1. Using $R = 68\ \Omega$, $C = 10^{-6}F$, calculate the transition frequency,

$$f_{tr} = \frac{1}{2\pi RC}.$$

2. Turn on the function generator. Set the frequency to 1 kHz and the amplitude to one-quarter of the maximum value.
3. Connect a BNC cable of the type shown in figure 18.13 to the function generator's output jack. Press the sine wave button.

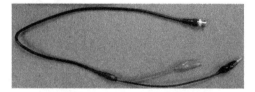

Figure 18.13. Recommended BNC cable.

4. Connect another BNC cable to CH1 of the oscilloscope.
5. Turn the oscilloscope and set the ground to the middle of the screen (figure 18.14).
6. Set the sweep time control to 0.25 ms/DIV and the vertical sensitivity for CH1 to 1 VOLT/DIV.
7. Next we will study the relationship between the pitch of a sound and the frequency. Insert the two wires from the headphone into two holes on the

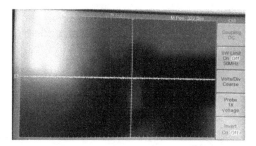

Figure 18.14. Ground set at the middle of the screen.

breadboard that are not connected underneath (with no conducting path underneath). Then connect the red and black leads of the BNC cables from the function generator and the oscilloscope to the headphone's red and black (white) leads (see figure 18.15). We should hear some sound from the headphone.

Figure 18.15. Connecting the BNC cables from the function generator and the oscilloscope and the wires from the headphone's on a breadboard.

8. Adjust the vertical sensitivity and the horizontal sweep time settings until we see a clear sine wave on the screen (e.g. Figure 18.16)

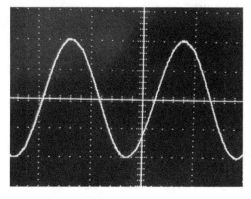

Figure 18.16. A sine wave

9. Determine the frequency of the sine wave from the period. Make sure we get a value close enough to the frequency displayed on the function generator. $f = \frac{1}{T} =$

10. Keep the oscilloscope settings the same, vary the function generator's frequency using the coarse frequency control knob, and see how the oscilloscope trace and the sound changes pitch. Make a careful observation and identify which sine waves in figure 18.17 (a) and (b) is a high-pitch or low-pitch and also which is a high or low frequency sound.

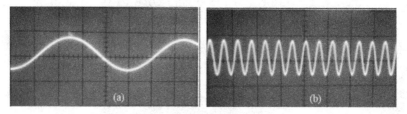

Figure 18.17. Oscilloscope trace for high-pitch and low-pitch.

Disconnect the red and black leads of the BNC cable from the headphone. Keep the BNC cables connected to the function generator and the oscilloscope.

18.4.4 High-pass and low-pass filters

We now verify that a resistor is high-pass and a capacitor is low-pass filters in an ac RC circuit. To this end, we need to build a circuit with a resistor, a capacitor, and an ac voltage (figure 18.18). We have built several similar circuits. Try to build this

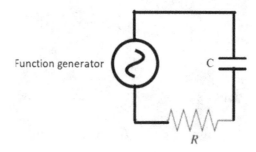

Figure 18.18. RC ac circuit design.

circuit without following instructions 1–5. All we need is the sine wave from the function generator, BNC cable, the resistor ($R = 68 \, \Omega$), the capacitor ($C = 1.0 \, \mu$ F), and the breadboard.

1. Connect the resistor and the capacitor in series on the breadboard.
2. Connect the red lead of the BNC cable from the function generator to the capacitor end **that is not connected to the resistor**.
3. Connect the black lead of the BNC cable from the function generator to the resistor end **that is not connected to the capacitor**.
4. Turn on both channels of the oscilloscope and set the grounds at the center of the screen. Make sure that the vertical sensitivity knobs for both CH1 and CH2 are set to 2V/DIV.
5. Connect channel one red and black leads of the BNC cable at the right places in the circuit to display the input voltage ($\Delta V_{in}(t)$). The input voltage is the ac voltage supplied by the function generator. (If you have difficulty, see figure 18.19.)

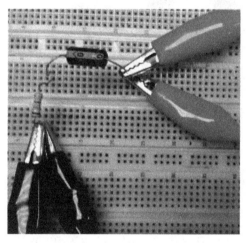

Figure 18.19. The red and black leads of the oscilloscope (channel one) and the function generator are connected to the resistor and capacitor which are connected in series.

6. When we calculated the transition frequency; we must have gotten, $f_{tr} \simeq 2340$Hz. Now set the input voltage frequency to the transition frequency by turning the coarse and fine frequency control knobs. Also, set the amplitude of the input voltage close to $\Delta V_{in,max} = 3$ V. We should see a trace like in figure 18.20.
7. Now connect another BNC cable to CH2 input of the oscilloscope and hook the red and black leads at the right places in the circuit to display the output voltage across the resistor ($\Delta V_{out}(t)$). (If you have difficulty, see figure 18.21.)

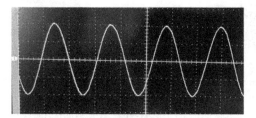

Figure 18.20. The input voltage.

Figure 18.21. Connecting the BNC cables from the oscilloscope (the two channels) and the function generator.

8. Block CH1 and display only CH2 ($\Delta V_{\text{out}}(t)$). Read the amplitude of the output voltage and estimate the uncertainties in reading.

9. Calculate the corresponding fractional uncertainty.

10. Calculate the experimental value for the gain across the resistor,

$$G_{\text{exp}} = \frac{|\Delta V_{\text{out,max}}|}{|\Delta V_{\text{in,max}}|}.$$

11. Calculate the theoretical value for the gain across the resistor using,

$$G_{\text{theo}} = \frac{R}{\sqrt{R^2 + \left(\dfrac{1}{2\pi f_{\text{tr}}\, C}\right)^2}}.$$

(18.20)

12. Is the experimental value close enough to the theoretical value within the limits of uncertainty? If it is, move to the next step.

13. What we will do next is pretty much the same as what we have done for the transition frequency. The only difference is that here we will take similar measurements for several frequencies to verify that a resistor is indeed a high-pass filter by doing a graphical analysis. So we need to read and record the values for the input voltage ($\Delta V_{\text{in,max}}$) and the corresponding output voltage ($\Delta V_{\text{out,max}}$) for the different frequencies listed on the table. The necessary steps we need to keep in mind are given below, a to c. Read $\Delta V_{\text{out,max}}$ and $\Delta V_{\text{in,max}}$ in divisions and volts for each of the frequencies listed in the table and calculate the gain, G. All the values must be recorded on the spaces provided in table 18.4.

Table 18.4. Measured frequency, input voltage and output voltage across the resistor.

Frequency (Hz)	$\log_{10} f$	$\Delta V_{\text{in,max}}$ (DIV)	$\Delta V_{\text{in,max}}$ (volts)	$\Delta V_{\text{out,max}}$ (DIV)	$\Delta V_{\text{out,max}}$ (volts)	G
10						
30						
100						
300						
1000						
3000						
10 000						
30 000						
100 000						
300 000						

(a) *Changing the frequency*: note that if you press the buttons on the function generator, for example, for the $1K$ range (this gives frequencies up to about 1 kHz) (figure 18.22). We can adjust the coarse and fine frequency control knobs until we get the required frequency.

Figure 18.22. Frequency range for the function generator.

(b) *Small frequencies*: for small frequencies you may get a flat line. We need to play with the horizontal sweep time setting knob to see a good sine wave trace on the oscilloscope. What we get could be an unstable trace. We do our best to find a way to read an accurate amplitude from this trace.

(c) *High frequencies*: you may get a trace that does not show picks of the sine wave for high frequencies. Again we need to play with the horizontal sweep time setting knob.

14. Using Excel, make a graph for the G vs f and sketch the graph in figure 18.23.

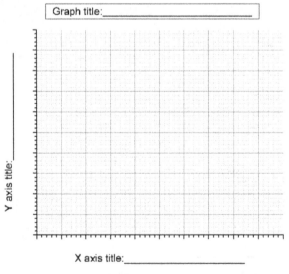

Figure 18.23. Graphing page.

15. Calculate the value $\log_{10} f$ and record the values in the space provided in table 18.4.

16. Using Excel, make a graph for the gain G vs $\log_{10} f$ and sketch the graph in figure 18.24.

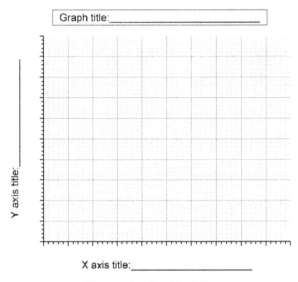

Figure 18.24. Graphing page.

17. We now proceed to verify that a capacitor is a low-pass filter. However, first, we will predict the behavior of the gain as a function of the frequency. To this end, we recall that theoretically, the gain for a capacitor is given by

$$G_{\text{theo}} = \frac{\dfrac{1}{\omega C}}{\sqrt{R^2 + \left(\dfrac{1}{\omega C}\right)^2}} = \frac{\dfrac{1}{2\pi f C}}{\sqrt{R^2 + \left(\dfrac{1}{2\pi f C}\right)^2}}. \qquad (18.21)$$

Using the resistor's ($R = 68\ \Omega$) and the capacitor's ($C = 10^{-6}F$) given values, find the gain for the two frequencies given in table 18.5.

Table 18.5. Measured frequency, input voltage and output voltage across the capacitor.

Frequency (Hz)	$\log_{10} f$	ΔV_{in} (DIV)	ΔV_{in} (V)	ΔV_{out} (DIV)	ΔV_{out} (V)	G_{theo}	G_{exp}
30							
30 000							

18. Using what we have learned in graphing G vs $\log_{10} f$ for the resistor and the values you obtained for G_{theo} and $\log_{10} f$, make a sketch that predicts the G vs $\log_{10} f$ for the capacitor in figure 18.25.

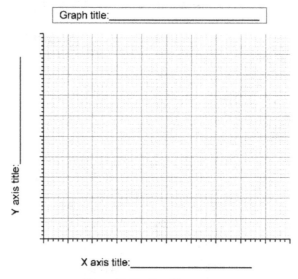

Figure 18.25. Graphing page.

19. Now, move the red and black leads of the BNC cable from CH2 and hook it across the capacitor to display the output voltage ($\Delta V_{out,max}$) in CH2. For the two frequencies listed in table 18.5, read the voltage amplitudes for $\Delta V_{in,max}$ from CH1 and $\Delta V_{out,max}$ from CH2 and record the values in the space provided on table 18.5.

20. Calculate the experimental values for the gain, G_{exp}, along with the experimental uncertainties.

21. Does the experimental values agree with the theoretical predictions we made, within the experimental uncertainties? If not, explain why?

22. Disconnect the BNC cable from the function generator and the circuit on the breadboard. Keep everything else just the way it was.

18.4.5 Filtering music

Next, we will study what a capacitor and a resistor do to music as low- and high-pass filters.

1. Turn the radio to your favorite music station. Keep the volume low.

2. Now, the ac voltage source would be the radio. So basically, we integrate the radio to the circuit as shown in figure 18.26. All we have to do is to connect the thin red wire from the radio to the capacitor and the other lead to the resistor on the breadboard (see how these leads are connected in figure 18.27). *Note: the thin red and white wires are from the radio.*

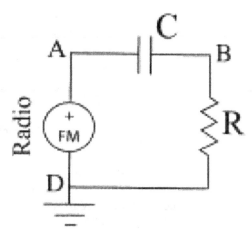

Figure 18.26. A radio connected to a resistor and a capacitor.

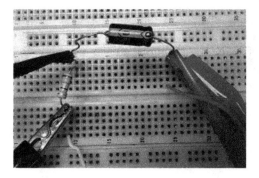

Figure 18.27. The radio connected to the circuit.

3. Channel one displays the signal from the radio (input signal), and channel two displays the output signal across the capacitor. You need to adjust the oscilloscope controls to get a good display. We may need to block one channel for a better observation.

4. Now, hook the red and black lead from channel two across the capacitor (see figure 18.27) and then across the resistor and observe what we see on the oscilloscope screen. We may need to block channel two. Remember, the capacitor is a low-pass filter, and the resistor is a high-pass filter.

5. After we figured out the main difference between the signals displayed for the resistor and the capacitor, make a rough sketch for each case that shows the difference in figure 18.28.

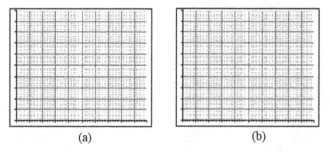

(a)   (b)

Figure 18.28. The signal across (a) the capacitor and (b) the resistor.

6. Now, connect the headphone across the capacitor (use the breadboard) to hear the signal passing through it. Do the same for the resistor. We repeat this step until we understand the difference in the pitches and frequencies of the sound we hear from the resistor and the capacitor.

18.4.6 Result and conclusion

1. In the G vs $\log_{10} f$ sketch for the gain curves for the high-pass (figure 18.24) and low-pass (figure 18.25) filters, label the attenuation, the transition, and pass regions. Briefly explain why the resistor is high- and the capacitor is low-pass filters in this circuit.

2. Write a brief overview of what we accomplished and concluded in this activity.

18.5 Real lab II: *resonance*

18.5.1 Objective

The objectives of this activity are

- To study an ac circuit with a resistor, an inductor, and a capacitor (RLC circuit).
- To measure input voltage, output voltage, and analyze gain across a capacitor in an RLC circuit.
- To study resonance.

18.5.2 Supplies

The supplies we need are alligators, cables, an oscilloscope, a function generator, BNC cables, headphones, a small radio, an inductor ($L = 10mH$), and a capacitor ($C = 1.0\mu F = 10^{-6}F$). Measure the resistance for the inductor and the capacitor and record the values.

18.5.3 Setting up the circuit and taking the measurements

We need to build a circuit (figure 18.29) with an inductor (which also has a resistance, as you recognized from your measurement), a capacitor, and an ac voltage in order to study resonance in an ac circuit.

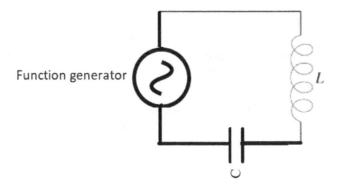

Figure 18.29. An ac RLC circuit design.

1. Turn on the function generator. Set the frequency to 1 kHz and the amplitude to the maximum value.

2. Connect a BNC cable to the output jack of the function generator. Press the button for the sine wave.

3. Connect the inductor and the capacitor in series on the breadboard.

4. Connect the red lead of the BNC cable to the inductor end that is not connected to the capacitor.

5. Connect the black lead of the BNC cable to the capacitor end **that is not connected to the inductor**.

6. Turn the oscilloscope on and set the ground at the time axis (middle of the screen) for CH1 and CH2. The vertical sensitivity knobs for both CH1 and CH2 to $2V/DIV$.

7. Connect another BNC cable to the CH1 input of the oscilloscope and connect the red and black leads to the right places in the circuit to display the input voltage ($\Delta V_{in,max}$) in CH1. The input voltage is the ac voltage supplied by the function generator. (If you have difficulty, refer to figure 18.19 in the previous activity.)

8. Now connect another BNC cable to the oscilloscope's CH2 input and hook the red and black leads to the right places in the circuit to display the output voltage across the capacitor ($\Delta V_{out,max}$) in CH2. (If you have difficulty, refer to the previous section.)

Virtual and Real Labs for Introductory Physics II

9. What we will do next is very much similar to what we have done in the previous section. We need to read and record the values for the maximum input voltage ($\Delta V_{in,\,max}$) and the corresponding output voltage ($\Delta V_{out,max}$) for different frequencies listed in table 18.6 and calculate the gain, G. If we have difficulties, we can refer to the previous activity.

Table 18.6. Measured frequency, input voltage and output voltage across the capacitor.

Frequency (Hz)	$\Delta V_{in,max}$ (DIV)	$\Delta V_{in,max}$ (V)	$\Delta V_{out,max}$ (DIV)	$\Delta V_{out,max}$ (V)	G
10					
500					
1000					
1100					
1300					
1350					
1400					
1500					
1550					
1700					
1800					
2000					
2500					
3000					
3500					
4000					
4500					
5000					

18-29

18.5.4 The resonance gain curve

1. Using Excel, make a graph for the gain (G) vs frequency (f). You should see a range of frequencies where the gain increases and then decreases. This is the region around the resonance frequency of your circuit.
2. Determine the resonance frequency f_{res} from the graph. If necessary, go back and take more data near the resonance frequency.

3. Sketch the graph in figure 18.30.

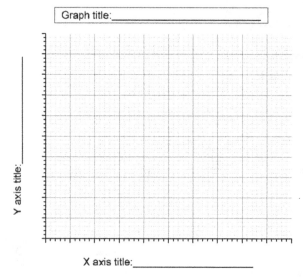

Figure 18.30. Graphing page.

4. The resonance frequency f_{res} is given by

$$f_{res} = \frac{1}{2\pi}\sqrt{\frac{2L - R^2C}{2L^2C}}.$$

From this equation, derive an expression for the resistance, R.
5. Using the result in step 4 and the resonance frequency f_{res}, the given values for L and C, find the resistance in the circuit.

6. What is the origin of the resistance in the circuit?

18.5.5 Result and conclusion

1. What is the dominant feature of the resonance curve we graphed? Relate this curve and its dominant feature to other resonance phenomena, such as pushing a child on a swing or a remote-control unit for a TV (remember that a charging voltage in a circuit means a charging electric field, and an electric field means a force on charges in the circuit). Therefore, an oscillating voltage means an electric force pushing back-and-forth on the charges in the circuit.
2. Write a brief overview of what we accomplished and concluded in this activity.

IOP Publishing

Virtual and Real Labs for Introductory Physics II
Optics, modern physics, and electromagnetism
Daniel Erenso

Chapter 19

Introduction to digital electronics

In this last chapter of the book, we conclude with an introduction to digital electronics. We begin with a basic introduction to Boolean algebra, which is the foundation for digital electronics. Then we carry out a real lab, with two parts, involving five integrated-circuit chips that include NOT, OR, AND, NOR, and NAND logical gates, a resistor, and an LED. In the first part, by building a circuit with each of these logical gates on a circuit breadboard, we verify the corresponding truth tables obtained from Boolean algebra. We then demonstrate the application of these logical gates by designing and building a bipolar light switch, which is a system of two light switches that both control a single light bulb.

19.1 Basic theory

The basic logic gates: the basic logic gates that we will study include: the NOT, the AND, the OR, the NAND (NOT AND), and the NOR (NOT OR).

- *One input gate*: the one input gate is the NOT logical gate. Suppose the input is A and the output is Y, the Boolean equation describing the NOT logical gate is

$$Y = \bar{A} \tag{19.1}$$

and the truth table 19.1

Table 19.1. Truth table for NOT gate.

A	$Y = \bar{A}$
0	1
1	0

19-1

- *Two inputs gates*: the AND, the OR, the NAND, and the NOR gates are two input gates with one output. If the inputs are A and B and the output is Y. The Boolean equations for these gates given in the table 19.2: the truth table for these three gates is given in table 19.3.

Table 19.2. Boolean equation for different gates.

Gate	Boolean equation
AND	$Y = A*B$
OR	$Y = A + B$
NAND	$Y = \overline{A*B} = \bar{A} + \bar{B}$
NOR	$Y = \overline{A + B} = \bar{A}*\bar{B}$

Table 19.3. Truth table for two input gates.

A	B	$A*B$	$\overline{A*B}$	$A + B$	$\overline{A + B}$
0	0	0	1	0	1
0	1	0	1	1	0
1	0	0	1	1	0
1	1	1	0	1	0

- *Reverse design*: forming the truth table from the circuit diagram.

19.2 Real lab A: *integrated-circuit chips*

19.2.1 Objectives

This activity will introduce us to five integrated-circuit chips (NOT, OR, AND, NOR, and NAND logical gates). For each of these chips, we will find the experimental truth table. We will identify the chip that is a NOT, OR, AND, NOR, or NAND logic gate from the experimental truth table (see figure 19.1). Measure the resistance of the resistor. It must be close to 560 Ω.

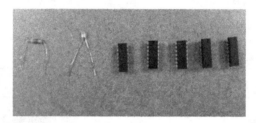

Figure 19.1. From left to right: a resistor, an LED, and the five chips.

19.2.2 Supplies

The supplies you need are the voltmeter, the breadboard, alligators, cables, shown in figure 19.1, small pieces of different colors of thin wires, a resistor, an LED, and five integrated-circuit chips.

19.2.3 Getting started

1. We will study the integrated-circuit chips by building each chip on the breadboard. What we have determined as the conducting and non-conducting paths of the breadboard is critical to build the chips on the breadboard and to take subsequent measurement. For additional help regarding the conducting paths of the breadboard refer to figure 19.2. Note that the red lines indicate holes connected underneath.

   For this activity as well as the next activity the power we need comes from the breadboard itself. What we need is a voltage close to 5 V and we can get that from the breadboard directly.

2. Turn the switch for the breadboard on. You must see a red light near the top of the switch. If you see the light turn it back off and move to the next step.

3. Set the top row of the breadboard to 5 V and the bottom row to the ground (0 V) by connecting red and black wires to the red and black leads of the breadboard (see figure 19.3)

Figure 19.2. The red lines show the conducting paths on the breadboard.

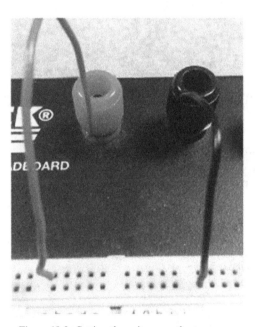

Figure 19.3. Setting the voltage on the two rows.

4. Turn the switch on. Measure the voltage in the first row and in the second row using the multimeter. In order to measure these voltages you must connect the black lead of the multimeter to the black lead of the breadboard. The red lead is inserted at the point where you want to measure the voltage (see figure 19.4 as an example for measuring the voltage in the first row).

Figure 19.4. Measuring voltage using the multimeter.

Voltage in the first row:

Voltage in the second row:

5. Turn off the switch. Remove the red lead of the multimeter from the breadboard but keep the black lead. We will use this method to check the voltage at different part of the breadboard for troubleshooting.
6. On the breadboard there are two columns that are vertically connected (see figure 19.2). We will use the two pairs in the middle of the breadboard. Set the first column to 5 V and the second column to the 0 V in each pair by connecting a wire from each of these columns to the two rows that you set the voltages in step 4 (see figure 19.5).

Figure 19.5. Setting the voltage along the two conducting columns.

7. For the left and right pairs of columns measure the voltage using the same approach discussed in step 4.

   Left pair:
   Voltage in the first column:

   Voltage in the second column:

   Right pair:
   Voltage in the first column:

   Voltage in the second column:

We have completed the ground work and we are ready to study the integrated-circuit chips. We study one one-input gate chip and four two-inputs gate chips (refer to the section 19.1). These chips have 14 pins which are numbered as shown in figure 19.6. Every chip that you use must have +5 V signal connected to pin 14, and the GND signal (0 V) connected to pin 7

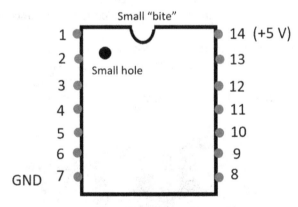

Figure 19.6. 14 pins chip.

19.2.4 One-input gate chip

The chip labeled 7404 is constructed from six gates each with one input and one output. The design is shown in figure 19.7. In this design the input gate is labeled A and the output gate is labeled Y. It is not hard to guess what kind of gates these are

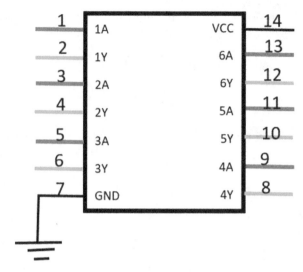

Figure 19.7. One-input chip design. Chip labeled 7404.

since we know only one gate with one input and one output. What would you predict this gate to be?

We now proceed to verify our prediction. We will do that by finding the truth table for only one of the gates, for example gate 2 (with input 2A and output 2Y). The truth table for the remaining gates will also be the same since all gates are the same in this chip.

1. Integrate 7404 chip to the breadboard by fully inserting the pins into the holes that are not connected horizontally. When you do that make sure the small 'bite' is facing upward! Set the 7th pin to the ground and the 14th bin to 5V (i.e. do exactly what is shown in figure 19.8).

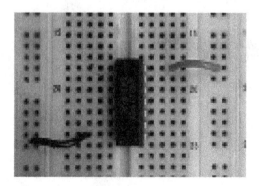

Figure 19.8. Setting the 7th and 14th pins voltage.

2. Now insert two wires in the rows of the breadboard where pin 3 (Input 2A) and pin 4 (Output 2Y) are inserted in (see figure 19.9).

Figure 19.9. Input and output wires.

3. Set the input wire (connected to 2A) to 0 V and measure the voltage at the output wire (connected to 2Y). Record the value in the table provided table 19.4.

Table 19.4. Measured input and output voltage for chip 7404.

Input voltage	Output voltage
0 V	
5 V	

4. Measure the output voltage when the input voltage is 5 V and record the value in the table provided.
5. Determine the logical values for the input and output voltages and record it in table 19.5.
6. Have you been able to confirm your prediction? If your answer is yes, proceed to the next step. *Note: chip 7404 is made of a NOT gate.*

Table 19.5. Logical values for input and output voltage for chip 7404.

Input voltage	Output voltage

7. A little bit farther below the chip, connect the resistor and the LED in series with the shortest leg of the LED grounded (see figure 19.10).

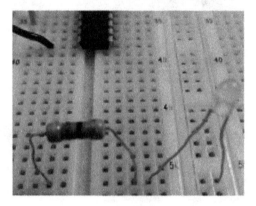

Figure 19.10. Resistor and LED in series.

8. Connect the output wire (connected to 2Y) to the end of the resistor (that is not connected to the LED). Turn the breadboard switch back on. By setting the input (2A) to 0 V and then to 5 V, find out when the LED is ON and OFF. Put your answer in table 19.6.

Table 19.6. Output voltage in terms of LED for chip 7404.

Input voltage	LED (ON or OFF)	Logical value
0 V		
5 V		

9. Put the logical values in the table when the LED is on and off.
10. Draw the digital circuit symbol corresponding to this logic gate with the input and outputs labeled (A and Y).

11. Explain why we did not connect the output directly to the LED. In other words why we added the resistor?

12. Turn the switch back off. Take out the input wire (connected to 2A) and the output wire (connected to 2Y). Leave the rest untouched!

19.2.5 Two-inputs gate chips

The chips labeled as 7408, 7432, 7400, and 7402 are constructed from gates with two inputs and one output. As you recall such kind of gates could be OR, AND, NAND, or NOR. We will identify these chips by finding the truth table for each of these chips.

Chips 7408 and 7432
The design for these chips is shown in figure 19.11.

Each of these chips have 2 gates on the left and 2 gates on the right sides. For example, on the left side pins 1-3 form one gate and pins 4-6 form another gate, which are the same type of gate. So we will determine the truth table for only one of the gates for each chips.

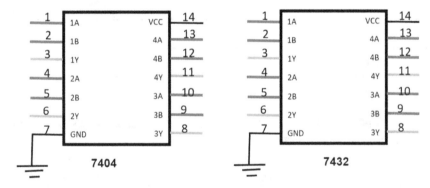

Figure 19.11. Two inputs chips. Chip labeled 7404 and 7432.

A. Chip 7408
1. Replace chip 7404 by chip 7408. Make sure the pins for chip 7408 are inserted in the exact same holes that the corresponding pins for chip 7404 were! and **the small 'bite' is facing upward!**
2. We are using gate 1 (with inputs 1A and 1B, output 1Y). Now insert three wires in the rows of the breadboard where pin 1 (input 1A), pin 2 (input 1B), and pin 3 (output 1Y) are inserted in.
3. Connect the output wire to the resistor in the same way you did earlier (see figure 19.12).

Figure 19.12. Two inputs and one output wires connected to the gate. The output is also connected to the resistor.

4. By turning the power back on and setting the input wires (1A and 1B) to the two possible voltage values (see table 19.7), determine if the LED will be on or off.

Table 19.7. Output voltage for chip 7408 in terms of LED.

Input 1A voltage	Input 1B voltage	LED (ON or OFF)
0 V	0 V	
5 V	5 V	
5 V	0 V	
0 V	5 V	

5. For results in table 19.7 find the corresponding logical values and identify the gate. Put your answers in table 19.8.

Table 19.8. Logical truth table for chip 7408 in terms of LED.

Chip # 7408 is _____ gate		
Input 1A	Input 1B	Logical value

6. Draw the digital circuit symbol corresponding to this logic gate with the inputs and the output labeled (*A*, *B*, and *Y*).

7. Turn the power switch off. Leave everything in place!

B. Chip 7432
1. Replace chip 7408 by chip 7432. Make sure **the small 'bite' is facing upward!**
2. Repeat step 3 and 4. But record your answer in table 19.9.

Table 19.9. Output voltage for chip 7432 in terms of LED.

Input 1A voltage	Input 1B voltage	LED (ON or OFF)
0 V	0 V	
5 V	5 V	
5 V	0 V	
0 V	5 V	

3. For the results in table 19.9 find the corresponding logical values and identify the gate. Put your answers in table 19.10.

Table 19.10. Logical truth table for chip 7432 in terms of LED.

Chip # 7432 is _____ gate		
Input 1A	Input 1B	Logical value

4. Draw the digital circuit symbol corresponding to this logic gate with the inputs and the output labeled (*A*, *B*, and *Y*).

5. Turn the power switch off. Leave everything in place!

Chips 7400 and 7402

Each of these two chips contains four identical, two-input logic gates, just as with the gates you investigated earlier. However, each of the gates contained on these chips is some combination of the gates that we already identified. The designs are shown in figure 19.13.

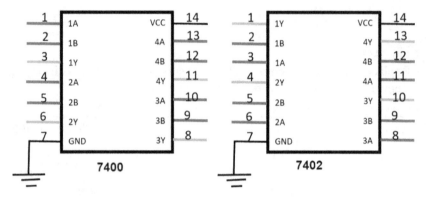

Figure 19.13. Two inputs chips. Chips labeled 7400 and 7402.

A. Chips 7400

1. Replace chip 7432 by chip 7400.
2. By turning the power back on and setting the input wires (1A and 1B) to the two possible voltage values, determine if the LED be on or off (see table 19.11).

Table 19.11. Output voltage for chip 7400 in terms of LED.

Input 1A voltage	Input 1B voltage	LED (ON or OFF)
0 V	0 V	
5 V	5 V	
5 V	0 V	
0 V	5 V	

3. For the results in table 19.11 find the corresponding logical values and identify the gate (table 19.12).
4. Draw the digital circuit symbol corresponding to this logic gate with the inputs and the output labeled (*A*, *B*, and *Y*).
5. Turn the power switch off.

Table 19.12. Logical truth table for chip 7400.

Chip # 7400 is _____ gate		
Input 1A	Input 1B	Logical value

B. Chips 7402

1. Replace chip 7400 by chip 7402.
2. For this chip pin 1 is output, pin 2 is input (input 1A), and pin 3 is input (input 1B) (see the design). Therefore, you must switch the role of the wires in pin 1 and pin 3. The wire going to the resistor must be moved from the row on the breadboard where pin 3 is inserted into the row where pin 1 is inserted in, while it is still connected to the resistor. Also move the wire that is connected to pin 1 to the row where pin 3 is inserted in. Now the two wires in pin 2 and pin 3 rows are the two inputs and you can call it 1A or 1B as you wish.
3. By turning the power back on and setting the input wires (1A and 1B) to the two possible voltage values, determine if the LED is on or off (table 19.13).

Table 19.13. Output voltage for chip 7402 in terms of LED.

Input 1A voltage	Input 1B voltage	LED (ON or OFF)
0 V	0 V	
5 V	5 V	
5 V	0 V	
0 V	5 V	

4. For the results in table 19.13 find the corresponding logical values and identify the gate (table 19.14).
5. Draw the digital circuit symbol corresponding to this logic gate with the inputs and the output labeled (*A*, *B*, and *Y*).
6. If you are confident that you have identified the four two-inputs logic gates (OR, AND, NOR, and NAND gates), turn the power off and dismantle what you built.

Table 19.14. Logical truth table for chip 7402.

Chip # 7402 is_____gate		
Input 1A	Input 1B	Logical value

19.2.6 Result and conclusion

Write a brief (3–5 sentences) overview of what was accomplished and concluded in this activity.

19.3 Real lab B: *digital circuit design*

19.3.1 Objectives

In this activity we will learn how to apply the skills in digital electronics that we developed so far to design, build, and test a digital circuit. We want to achieve the following objective:
1. define appropriate binary variables and build a truth table for desired function of the circuit,
2. draw the circuit diagram from the truth table using the minterms and the Boolean equation,
3. build the circuit from the diagram, and
4. test it to reproduce the truth table and verify if it functions properly.

19.3.2 Supplies

The supplies you need are the breadboard, pieces of thin wires, a resistor, an LED, and the five integrated-circuit chips.

19.3.3 The challenge

A bipolar light switch is a system of two light switches that both control a single light bulb. (For example, one switch can be at the bottom of a set of stairs and the other at the top; they both control the light above the stairs.) The light switches we will consider are the type that switch up and down.

Let's say that the two switches start off in the down position, and the light is off. The rule is that, if either light switch is switched at any time, the state of the light bulb changes. For example, with both switches down, we know that the light is off. If the switch at the top of the stairs is switched up, then the light will go on. If the other switch at the bottom of the stairs is then also switched up, then the light will go off.

Our challenge is to design, build, and test a digital circuit using the three fundamental digital circuit chips we studied in the previous section (NOT, AND, and OR gates) that represent a *bipolar light switch*. The *output of the circuit should feed into an LED that represents the light bulb.*

19.3.4 Designing the circuit

1. We define the following symbols for the inputs and the output variables (table 19.15):
2. We also assign the following binary (logical) and voltage values when the switch is up or down and the LED is ON or OFF (table 19.16)
3. Using the symbols defined and the assigned binary values, construct the truth table for the circuit we need to build that serve as a bipolar switch (use the table 19.17).

Table 19.15. Symbols for inputs and output.

Variable	Symbol
Input 1 (Switch 1):	A
Input 2 (Switch 2):	B
Output (LED):	Y

Table 19.16. Binary values and voltage.

	Binary (logical) value	Voltage value
Switch up	1	5 V
Switch down	0	0 V
LED ON	1	3 – 5 V
LED OFF	0	0 V

Table 19.17. Circuit truth table.

A	B	Y

4. In the truth table identify the rows that have the minterms. Using the minterms determine the Boolean equation.

   The Boolean equation:

5. Using the appropriate logic symbols draw a circuit diagram for the Boolean equation in the space provided.

19.3.5 Building and testing the circuit

1. Construct the digital circuit from your circuit diagram. Here you will not be provided step by step instruction to build the circuit you designed. You just need to apply what you were introduced to the different logic gates in the previous section. However, we will provide you with following basics:

    (a) According to your circuit design, you should realize by now that you need a NOT, an AND, and an OR gates. You must identify the chips with these gates from the previous activity and build it on the breadboard (see figure 19.14). Note that **the small 'bite' for each chip must be facing upward on the breadboard!**. Also the chips on the breadboard should be build in the following order: NOT, AND, and OR gates.

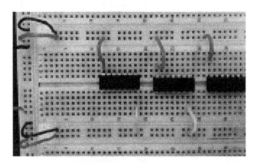

Figure 19.14. The three chips.

    (b) You must build the resistor and the LED connected in series near the chip with the OR gate with the short leg of the LED grounded (see for example figure 19.15)

    (c) You must define your inputs (A and B) going into the NOT gates (see for example figure 19.16)

    (d) You must have a wire from the output of the OR gate to the resistor (see for example figure 19.17)

    (e) You are recommended to use the gates either on the left side or the right side for all chips.

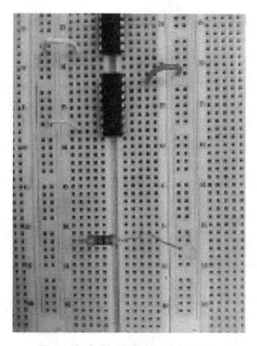

Figure 19.15. The resistor and the LED.

Figure 19.16. The two input wires.

2. After you completely constructed the digital circuit, test if it satisfies the truth table from step 3 of Designing the circuit.

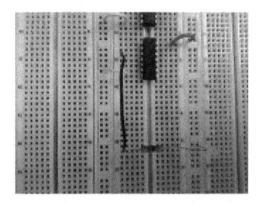

Figure 19.17. The OR gate connected to the resistor.

19.3.6 Result and conclusion

Write a brief (3–5 sentences) overview of what was accomplished and concluded in this activity.

CPSIA information can be obtained
at www.ICGtesting.com
Printed in the USA
BVHW060830190722
642202BV00002BB/12